BIZARRE WEATHER

Bizarre Weather

Howling Winds, Pouring Rain, Blazing Heat, Freezing Cold, Hurricanes, Earthquakes, Tsunamis, Tornadoes, and More of Nature's Fury

JOANNE O'SULLIVAN

An Imagine Book
Published by Charlesbridge Publishing
85 Main Street, Watertown, MA 02472
(617) 926-0329
www.charlesbridge.com

Library of Congress Cataloging-in-Publication Data

O'Sullivan, Joanne.
Bizarre weather : howling winds, pouring rain, blazing heat, freezing cold, huge hurricanes, violent earthquakes, tsunamis, tornadoes and more of nature's fury / Joanne O'Sullivan ; illustrated by Jeff Albrecht.
p. cm. — (An imagine book)
Includes bibliographical references.
ISBN 978-1-936140-72-5 (alk. paper)
1. Climatic extremes—Juvenile literature. 2. Storms—Juvenile literature. 3. Droughts—Juvenile literature. 4. Natural disasters—Juvenile literature. I. Albrecht, Jeff, ill. II. Title.
QC981.8.C53O88 2012
551.55—dc23

2012016236

Cover and interior design by Cindy LaBreacht. Illustrations by Jeff Albrecht.

Printed in China, August 2012

2 4 6 8 10 9 7 5 3 1

ISBN 13: 978-1-936140-72-5

For information about custom editions, special sales, premium, and corporate purchases, please contact Charlesbridge Publishing at specialsales@charlesbridge.com

Contents

Introduction

On a clear day in April 2011, a group of kids met up to play soccer after school at Galashiels Academy in Scotland. They had just started warming up when they heard strange thudding sounds. Alerted by the unusual noise, their gym teacher looked around and noticed that there were worms writhing around on the ground near the students. Then the kids started to cover their heads. The worms were falling from the sky—more and more of them with each passing second. The incident lasted just a few minutes, but when it was over, hundreds of worms had dropped to the playing field and onto the tennis court, according to the teacher, who collected some of the worms as samples. The playing field wasn't close to any buildings on campus, eliminating the possibility that the group had fallen victim to a prank. That left just one possibility: this soccer team had just experienced some bizarre weather.

Horses are carried aloft by tornadoes and set down unharmed. Hailstones as big as elephants plummet to the earth. There are whirling columns of fire, clouds *created* by airplanes, and places where lightning rarely stops flashing. Into the furthest reaches of recorded history, there are stories of strange and unbelievable weather, which was at the time chalked up to witchcraft, devilry, and the end of the world. (When the sky turned pitch black in the middle of the day, surely the end was near!). Our ancestors had no way of knowing that those dark days were just the result of ash particles from raging forest fires or volcanoes mixed with a thick fog.

As technology and forecasting have improved, scientists have been able to offer explanations for some of the most puzzling mysteries in weather history. On the other hand, there are phenomena that still can't quite be explained. And technology can help us uncover even *more* baffling phenomena, such as high-atmosphere, colored lightning and cloud systems that rain in unison.

There's nothing like the weather to remind us all that we live in a world that's not always predictable and rational. No matter how much we believe we have conquered nature, it will still have a lot of surprises in store for us. Even in the twenty-first century, a dust storm can shut down a major port city, black rain can pour from the sky, and cities in warm climates can be buried in ice. And, yes, schoolchildren can be pelted with worms.

Forget what you think you know about the weather. The stories in this book will shock, surprise, amuse, horrify, and puzzle you. Collected from historic records, present-day news stories, and research studies, these accounts of bizarre weather capture the volatile and changeable nature of everyone's favorite topic. While you may start out a skeptic, you'll find that bizarre weather does happen, even if hasn't happened to you—*yet*. Somewhere in the world, at this moment, the rain may be red, the wind vicious, and the temperatures unbearable. Expect the unexpected, and keep your eyes peeled for airborne invertebrates.

Peculiar Precipitation

Most of us never experience anything more out of the ordinary than a particularly heavy downpour or perhaps a snowstorm big enough to send everyone out to the grocery store for bread and milk. While unexpected, these weather events don't fall into the realm of the mysterious or truly bizarre. But when the rain starts changing colors or things fall from the sky that have no business being in the sky in the first place, it really gets our attention. Frogs—or even what appears to be blood—dropping from the clouds are the stuff of legend and biblical prophecy. Surely they're portents of doom or maybe even the end of the world. After all, when the end comes, it seems we're all expecting it to come from *above*. In reality, there's usually a perfectly scientific explanation for these strange occurrences. And when there's not, scientists will continue to search for one.

Rainy Day Blues (and Reds, Blacks, and Yellows)

The color of raindrops isn't always crystal clear. While there don't seem to be any stories featuring rain in shades of gray, you'll find that explanations about the causes and sources of colored rain can often be ambiguous.

RED RAIN

As early as Roman times there have been reports of "blood rain." This red-tinted rain has often been interpreted as an omen or sign of the gods' displeasure. In fact, most recorded "blood rains" have been in Europe and are the result of sand and dirt particles from the Sahara Desert being borne up into clouds and redistributed in raindrops.

Charles Darwin wrote about a red rain that he experienced while at sea off the Cape Verde Islands. Darwin estimated that the rain extended for an area of 1,600 miles and reached a distance of up to 1,000 miles off the African coast.

In 1827, there were reports of "blood rain" falling near the village of Buliavino, Russia. Farmers reported that after the rainstorm, the fields of the village turned orange and then blackened as if covered by soot. Later, residents claimed that their vegetables had grown far beyond their normal size, and people of the village were healthier than ever before. Some who had saved the rain in jars started using it as an elixir for vitality or began selling it as such.

In 1860 in Siena, Italy, a red rain lasted for two hours, and then was followed by two more red rainfalls the same evening. Just a few days later, there was another red rain. When it started, observers said, the rain looked like watered-down red wine. By the time of the last shower, there was only a slight reddish tint to the rain.

The best-known red rain of modern times was the red rain of Kerala, India, in 2001, which continued falling over the course of two months.

Early reports blamed the rain on the explosion of a meteor. Later theories included dust, volcanic ash, and algae spores. Environmental groups attributed it to acid rain, a type of precipitation tainted by pollution from industrial sources. Years after the event, a prominent scientist proposed that the cells found in the red rain's particulate matter were extraterrestrial. Beside the E.T. rain rationale, no definitive explanation has ever been offered.

YELLOW RAIN

You've heard the expression "don't eat yellow snow." But you've probably never heard of yellow *rain*. There were reports of just such a thing in Kentucky in 1867 and in Germany in 1886. During those incidents, witnesses reported that the rain smelled like sulfur, and since that substance was associated with the Devil, those downpours were called "the Devil's rain." Although most so-called "sulfur showers" are actually just the result of excessive pollen in the air (see page 12), there have been a few incidents of actual lumps of sulfur falling from the sky. In 1867, residents of Thames Ditton, Surrey, England, reported seeing a rain of fire that lasted about ten minutes. The next morning they found thick deposits of sulfur in water pools outside. In 1873 in Hungary, sulfur lumps fell in an area surrounding the site of a fallen meteorite.

Not long after the disastrous earthquake and tsunami that took place in Japan in 2011, yellow rain fell in the Kanto Prefecture. Local residents were concerned that it might be connected to the nuclear reactor that had been compromised in the Sendai region. After all, yellow rain had fallen in Chernobyl, Ukraine, after the nuclear disaster there. However, that country's chief meteorological agency discovered that the yellowy substance was just pollen.

Pollen was also cited as the cause of the yellow rain that fell in the Bellarine Peninsula in Victoria, Australia, in 2011. Authorities warned residents that while their allergies might act up, there was no reason to be alarmed.

BLACK RAIN

It's not uncommon for a volcanic eruption to cause what's referred to as "black rain," which occurs when ash particles get swept up into clouds and are redistributed during rain showers. That phenomenon also occurs when there's been a forest fire in the area. In 1846 in Worcester, England, there were reports of rain that blackened the clothing and skin of farmers who were working outdoors when it fell. Those rains were attributed to particles from nearby coal mines. Around the same time in England, actual lumps of coal were said to have rained from the sky. The assertion was also made in Kentucky in an area close to a coal mine.

In modern times, black rain can also be caused by industrial pollution. Shenzhen, China, is known for its black rains. The rain leaves black residue on clothing, umbrellas, and cars. It's so acidic, it's even been known to eat through the petals of flowers, leaving holes in them. Local authorities traced the source of the black rain to the nearby Nanshan Electricity Plant, which burns heavy oil to generate electricity, and then releases the ash residue from the furnace into the air. Power plant officials say the black rain is unavoidable, and they try to schedule the ash release at night so fewer people will be affected. For decades, the same region has experienced both black rains and black snows, with some of the incidents occurring 160 miles to the north and south. In 1994, the black rain in Chongqing was ink-black in color and highly corrosive, the result of sulfur mixed in the coal that was being burned.

"Cinder falls" have been relatively common since the Industrial Revolution. In 1939, coal is said to have fallen in Springfield, Missouri. In 1983, in Dorset, England, lumps of coal fell during a heavy storm.

GREEN RAIN

A rare green rain fell in India in 2002. Local people feared that it was the result of a chemical weapons attack, but after the chief pollution scientist in the area examined it, it was revealed to be bee feces, which were full of pollen from mangos and coconuts. Wide-scale occurrences of this "green rain" could happen during the migration of Asian honeybees.

MUD BATH

Mud falling from the sky is not as unusual as you would think. An 1879 mud shower in Nevada dropped mud so thick that it stopped a passenger train in its tracks. Muddy rain fell across an unusually large area of the U.S. (including Pennsylvania, New Jersey, New York, and Illinois) in 1902. Mud rain isn't that uncommon in Texas, and it's becoming something of a frequent occurrence in Beijing, where rain mixes with sandstorms coming from Inner Mongolia. Mud rain has also been reported in the Caucuses, the result of sandstorms blowing in from the Arabian Peninsula.

Beyond Cats and Dogs: Falling Fish, Frogs, Snails, and Rats

The accounts of frogs, snails, fish, and even snakes falling from the sky during rainstorms are isolated and sometimes hard to believe. For centuries, people thought the tellers of these tales were anywhere from slightly dotty and daft to downright crazy. But as scientific evidence has become more available (not to mention firsthand photo and video accounts), it's clear that the clouds do open up and rain down a variety of creatures, both dead and alive. Some people accept this as a sign of divine abundance and either keep the animals or cook them for dinner.

FISH TALES

By far the most frequently cited type of cloud-bursting beast is the fish. A scientist at the American Museum of Natural History compiled seventy-eight reports of fish falls over the course of decades, but many have occurred since his research concluded in the 1950s. Fish falls have been reported all over the world, but typically tend to happen in areas close to rivers or the ocean.

Accounts of fish falls differ. But usually witnesses report first hearing a thud or thudding sounds before discovering that fish are falling from the sky or flopping on the ground around them. The most common explanation for fish falls such as this is that the fish are transported into the clouds via a waterspout, which is a vortex that forms over water. (Some describe them as water tornadoes. See page 68.) Skeptics counter that this explanation doesn't account for fish that fall to the ground dry, dead, or headless.

Fish fall from the sky so regularly during the months of May and July in the Honduran region of Yoro that local people hold a festival to celebrate their arrival. It's known as *Festival de la Lluvia de Peces*: the Rain of Fish Festival. Because the fish falls—which started around

the 1970s—are said to provide people with hundreds of fish to take home and cook, the festival features a lot of grilled fish. While some speculate that the fish falls are due to waterspouts from the Atlantic coast about eighty-seven miles away, others have suggested that the fish might come from freshwater rivers nearby and emerge from underground springs rather than fall from the sky. True believers attribute the fish falls to divine intervention, saying that the falls are miraculous. In the nineteenth century, the area was home to a priest who was devoted to alleviating poverty. The faithful today believe that prayers directed to him are responsible for the fish falls. As a result, he has been nominated for sainthood.

One of the best-known incidents of a fish fall happened in Marksville, Louisiana, in 1947, when hundreds of fish rained down on the city, landing on pedestrians as they went to work. A wildlife biologist who happened to be in the area on assignment for the Department of Fish and Wildlife recalls that he was eating breakfast in a local diner when the waitress told him that fish were falling from the sky. He immediately went out to collect specimens. He found they were "freshwater fish from local waters," including the large-mouth bass, two species of sunfish, minnows, and hickory shad, ranging in size from two to nine inches. The biologist reported that fish had fallen at a rate of one per square yard along the town's main street, evenly covering the area. What's more, the fish were "perfectly fresh and suitable for human consumption," giving the residents a one-day rest from fishing for their supper. Speculation about the occurrence immediately turned to waterspouts since the town is not far from the Mississippi River. However, no waterspouts had been reported in the area.

A 1986 "smelt storm" on Lake Huron in Canada was reported by fishermen on the *Rhonda K*, who said that thousands of the small fish fell all around their boat. Attracted by the fish, hundreds of gulls appeared, picking the fish out of the air.

FISH FALL TIMELINE

1830: Feridpoor, India. Witnesses reported seeing something in the sky that looked like a flock of birds. It started to move closer to the ground, at which point the witnesses say they were pelted with up to four thousand fish, some fresh, some dead, dry, and headless. Of note, said a British physician who was present, was the size of the fish: from one-and-a-half to three pounds each.

1859: Glamorganshire, Wales. On an "uncommonly wet" day, a man reported two separate fish falls of roughly two minutes each, occurring with a ten-minute interval in between. The fish fell in an oblong-shaped area about eighty yards long by twelve feet wide. The fish were still alive and the man and his friends scooped up buckets full of them. This became known as the Great Stickleback Fish Fall. (Stickleback is the type of fish that fell.)

1861: Singapore. After an earthquake, local residents reported a fish fall.

1896: Essen, Germany. *Frozen* fish fell from the sky.

1906: Coopers Plain, Australia. Hundreds of live fish fell from the sky.

1909: Newcastle, South Africa. A fall of hundreds of small fish was reported, with some alive and some dead.

1918: Hindon, England. During a thunderstorm, hundreds of small fish, identified as sand eels, fell from the sky. All the fish were dead and already stiff.

1936: Guam. An American army officer stationed on the island reported a shower of fish. The fish fall was notable because the fish were typically only found in waters near Europe.

1948: Bournemouth, England. Herring fell on a golf course.

1956: Chilatchee Creek, Alabama. A couple reported they observed a cloud form and turn dark. They watched as it released a rain of live catfish, bass, and bream before turning white again.

1984: Los Angeles, California. A fish fall over the Santa Monica Freeway disrupted traffic.

1989: Ipswich, Australia. Around eight hundred sardines fell on a couple's yard.

2000: Ballyconnell, Sligo, Ireland. After a heavy rainstorm, a man discovered silver fish flopping around outside his cottage. He assumed that they had been dropped by sea gulls, but later found some on the roof, and then suspected that they had come in with the storm.

2010: Lajamanu, Australia. Residents of this remote desert outpost reported that hundreds of spangled perch fell from the sky. Most were alive in the air, but died as they hit the ground. It was the second time the town had been the scene of a fish fall, the first being in the 1980s. The town is nearly four hundred miles from the coast. A nearby tornado was blamed for the more recent fall.

2011: Florida. In two separate incidents, a single fish was said to have dropped from the sky, each landing on cars in different areas of Florida. The areas were far enough inland to eliminate the possibility of them being flying fish.

FLYING FROG TIMELINE

Frogs seem to be the second most likely creature to rain down from the sky. Perhaps because of their biblical association with the plagues, frog falls tend to promote panic.

1794: Lalaine, France. Members of the French army reported toads falling from the sky during heavy rain.

1804: Toulouse, France. Hundreds of baby frogs fell from the sky.

1838: London, England. Small frogs and tadpoles dropped into the city center during a severe storm.

1873: Kansas City, Missouri. A thick layer of frogs fell during a thunderstorm.

1901: Minneapolis, Minnesota. A heavy storm brought down a giant green mass from the sky. When the storm was over, residents found a four-block area covered with frogs and toads. In some places, the piles of amphibians were so deep that it was impossible to move through them.

1939: Trowbridge, England. Witnesses reported hundreds of frogs dropped from the sky.

1969: Buckinghamshire, England. Dozens of witnesses reported a frog rain.

1977: Morocco. A rain of frogs was reported in the Moroccan Sahara.

1981: Nafplion, Greece. A rain of tiny frogs, each weighing just a few ounces, was reported in this seaport town. The frogs were later found to be native to North Africa.

2009: Ishikawa Prefecture, Japan. Dead tadpoles fell from the sky over the course of a month in an area spanning several miles.

WORM FRONT MOVING IN

Besides the Scottish worm fall story that began this book, there was also a report of worm rain in Jennings, Louisiana, in 2007. A woman saw clumps of live worms falling from a cloudless sky. Although no cause was ever confirmed, speculation pointed to a waterspout that had been seen just five miles away around the same time. Prior to that time, incidents of worm rains had been reported in Sweden in 1924, Norway in 1877, and Massachusetts in 1872.

FALLING FAUNA TIMELINE

1680: Germany. Rats are said to have fallen from the sky during a thunderstorm.

1857: Montreal, Quebec. Lizards are reported to have fallen from the sky.

1869: Chester, Pennsylvania. During a rainstorm, snails fell to the ground in a slow, whirling motion.

1877: Memphis, Tennessee. Thousands of snakes, ranging in size from twelve to eighteen inches, appeared on the sidewalks after a heavy rain. No one actually saw them fall.

1877: Charleston, South Carolina. Two very small alligators were thought to have fallen from the sky during a storm as a result of a nearby waterspout.

1880: Owensville, Kentucky. A "veritable shower" of beetles fell on the town's residents. The beetles were thought to have been migrating when caught up in a strong wind.

1881: Worcester, England. Local residents experienced a shower of periwinkles and crabs.

1890: Messignadi, Italy. Bird blood was reported to have fallen during a violent windstorm. Locals speculated that birds had been torn apart by the wind. However, it may have just been red rain (see page 10).

1892: Paderhorn, Germany. Hundreds of mussels fell from an unusual yellowish cloud.

1930: Danville, Virginia. After a thunderstorm, seashells were found on the ground a hundred miles from the nearest shore.

2007: Salta Province, Argentina. A variety of spiders, some as big as four inches in diameter, fell from the sky during a rainstorm.

THE BIRDS

Is the end of the world at hand? Several separate bird fall incidents around the beginning of 2011 caused many to believe so. One massive bird fall, totaling approximately five hundred redwing blackbirds, took place on New Year's Eve in Louisiana. At almost the same time, five thousand blackbirds slammed into roofs in Arkansas. One hundred birds were found dead in the streets in Sweden during the same period, along with eight thousand turtledoves in Italy.

While the media spread alarm, scientists stated that such incidents are not that unusual. In the case of the birds in Louisiana, the deaths were attributed to shock from hitting live power lines. In the other locations, they were victims of shock thanks to New Year's fireworks. The birds in Italy simply died from indigestion and overeating.

Grainy Rain

It's not just animals that fall from the sky. Agricultural products of many varieties have been known to shower down from the heavens. Just as biblical manna was said to be sent for those who needed food, there are isolated incidents of edible substances falling from the sky.

FALLING FLORA TIMELINE

1857: Lake County, California. One-quarter-inch (and smaller) sugar crystals fell from the sky.

1867: Macereta, Italy. A number of "blood red" clouds appeared in advance of a cyclone. The storm then scattered seeds all over town, covering the ground in a half-inch layer. The seeds were found to be from the Judah tree, a species that grows in North Africa.

1867: Dublin, Ireland. A shower of "aromatic-smelling berries," black in color and about one inch in diameter, fell from the sky so swiftly and with such great speed that several people were injured by them.

1879: Falkland, Scotland. After a heavy hailstorm, seaweed was found hanging from trees and on the ground across the countryside. A waterspout had been reported nearby.

1890: Turkey. A substance the locals referred to as manna fell from the sky in rural Turkey. Residents collected it and made it into bread, finding it to be very good to eat.

1961: Shreveport, Louisiana. Small green peaches fell from the sky.

1963: Kent, England. Straw fell from a clear sky during an hour-long "straw storm."

1977: Devizes, England. A "cloud" of hay drifted over the town and then fell in small clumps.

1977: Bristol, England. A couple walking down the street reported a shower of fresh hazelnuts. Another person reported the same phenomenon in the same spot just a few minutes later. Each witness said that there were no hazelnut trees in the vicinity.

1982 to 1986: Evans, Colorado. Falls of corn kernels were reported on numerous occasions in the area during this period.

1992: Basildon, Essex, and South Wonston, Hampshire, England. Balls of hay rained down on both towns on the same day.

1994: Buckinghamshire, England. Clumps of hay, "some as big as bales," fell from the sky. The incident was attributed to the fact that it was haying season in the region and the hay had been taken up in a whirlwind and deposited elsewhere.

2001: Wichita, Kansas. Dried cornhusks were reported to have fallen over the downtown area.

2009: Elberton, Georgia. Hay fell from a sunny sky in wisps approximately eight inches long.

Gory Storms

It sounds like the stuff of a horror story—flesh and fur falling from the sky. While such occurrences are by no means common, they've happened more than you'd think possible. (There seems to be a perplexing number of this type of incident in California in the nineteenth century.) Explanations for this phenomenon are few and far between. Was the airborne meat dropped by flocks of birds as they passed overhead? Or, could there have been a particularly violent storm that tore apart the bodies of animals and deposited them elsewhere? Regardless of the causes, these gory storms are among the most bizarre (and disgusting) weird-weather events recorded.

GORY STORM TIMELINE

1841: Lebanon, Tennessee. A tobacco farm was pelted with animal blood, fat, and muscle tissue, which witnesses say fell from a red cloud. A rattling noise was heard as the material issued from the cloud.

1850: Benicia, California. Soldiers at an army base reported being struck with pieces of meat falling from the sky in a downpour that lasted three minutes. The meat came in a variety of shapes and sizes, ranging from the size of "a pigeon's egg to that of an orange." The heaviest weighed three ounces. The fall covered an area of ground approximately nine hundred feet long by four hundred feet wide. When the surgeon on base examined the flesh, he determined it to be beef, and found some blood vessels and muscle in the specimen. It was not completely fresh. There was a brisk breeze in the area, but nothing more unusual was reported. The story was recorded in the *San Francisco Herald*.

1863: Sacramento, California. A meat shower was reported to have fallen on a single farm on the edge of town. The incident, which was witnessed by two people, began when they noticed that the farmer's chickens were pecking at something on the ground. Upon closer observation, the witnesses saw that it was flesh, and they then noticed that it was falling around them from the cloudless sky. The farmer collected some samples before the chickens could eat all of them, noting that the samples were three to four inches long and approximately two inches wide, each weighing approximately one-quarter pound. The composition of the specimens appeared to be liver mixed with grain.

1869: San Jose, California. Witnesses reported meat, bone, blood, brains, and "nerves" falling from a clear sky over an area of approximately twenty acres. The material had a fishy smell. They collected samples and brought them to the San Francisco Science Academy for study.

1869: Los Nietos Township, California. While attending a child's funeral, a group of stunned mourners were horrified when they were caught in a shower of meat and blood. The meat was mixed with animal fur, and ranged in size from fine particles to ten-inch chunks. The shower fell over an area of approximately two acres.

1870: Riverside County, California. Clumps of coagulated blood were reported to have fallen from a cloudless sky over a farm. Local news reports stated that the blood splattered the doorstep and the surrounding grounds. Witnesses said that a whirlwind raced across the property just before the blood was discovered.

1871: El Monte, California. Raw, but "boiled-looking," meat was reported to have fallen on a farm, and was said to "drench" the farm's cornhusks with blood.

1876: Olympian Springs, Kentucky. A woman reported a shower of meat falling in an area about one hundred yards long and fifty yards wide. She said the sky was clear at the time when something resembling large, dark snowflakes started to fall. She then realized it was flakes of fresh meat approximately three to four inches square in size. The meat shower lasted about two minutes. Two men tasted the meat and judged it to be either mutton or venison. The story made the headlines of the *New York Times* on March 10, 1876.

Random Rains

These rainfalls defy both explanation and categorization.

ODD BALLS

In 1781, a farmer in Lyon, France reported that a rock fell through his thatched roof and cracked the oak table where he was sitting. It split open in front of his eyes, revealing yellow crystals that smelled like rotten eggs and later turned black. When examined by experts in Paris, it was determined to be brimstone, otherwise known as the Devil's Stone.

Golf balls are said to have rained from the sky in Punta Gorda, Florida, in 1969.

A shower of ball bearings fell from the sky in Greece in 2008.

A strange metallic object, said to the be size of "a baby's fist," fell from the sky in Freehold Township, New Jersey, in 2007. It punched a hole in the roof and broke bathroom tiles before becoming embedded in a wall. Investigators weren't able to determine the nature of the object.

CLOUDLESS RAIN

While it's a fairly frequent occurrence, there's no single explanation for rain that falls from a seemingly cloudless sky. The most common speculation is that the sky isn't really cloudless. Wind can carry rain over significant distances and a vertical wind shear can make it appear as though rain is falling from overhead. High-altitude clouds, which can't easily be seen, may also be carrying the rain horizontally. Another theory is that the "rain" people experience on cloudless days is actually a kind of fog.

In 1881, the *New York Times* reported that in Chesterfield County, just outside of Charleston, South Carolina, rain fell from a cloudless sky every day between the hours of 9 a.m. and 1 p.m., in an area covering just a few blocks.

HIGH AND DRY

Rain that falls from the sky doesn't always reach the ground. This is called *virga*, and is quite common in places that are very dry.

SPARKLING RAIN

During an 1892 thunderstorm in Cordova, Spain, witnesses reported that a flash of lightning ripped the sky, and, immediately afterward, the drops of rain made a crackling sound and emitted sparks as they hit the ground.

BIG DRIPS

Some of the largest raindrops ever recorded were discovered by scientists flying in planes through cumulus clouds over Brazilian rain forests and the Marshall Islands. The drops—described as parachute or jellyfish-shaped—were eight millimeters to one centimeter in diameter. Scientists speculate that the large size was the result of water condensing around ashes from a fire in the Brazilian rain forest below, and, in the case of the Marshall Islands, around salt from the ocean's water. Raindrops that size wouldn't be found on or near the ground because they typically break up when they collide with other drops on their way down to the earth.

LUCKY DAY

In 2011, police in the Dutch city of Maastricht reported that a transport truck carrying money broke open. For a few moments, motorists reported that it was "raining" euro notes. Drivers pulled over and tried to catch the notes, causing traffic snarls.

In 1976, two German clergymen were the recipients of two thousand marks that floated down from a clear sky in Limburg.

In 1940 in Russia, silver coins are said to have fallen from the sky during a rainstorm.

Snow Weird

Snow can be just as unpredictable as rain, resulting in such strange phenomena as snow donuts, blood snow, snow cones, and rainbow snow.

SNOW IN STRANGE PLACES

The first snow ever recorded in the lower elevations of the Sahara Desert fell in the town of Ghardaïa, Algeria, on February 18, 1979. Although Saharan mountain ranges (such as the Tibesti Mountains in Chad or the Ahaggar Mountains in Algeria) receive snow about every seven years, this February snowfall was the first time anyone

could remember it falling outside of the mountains. The half-hour snowstorm put the city at a standstill, but the snow melted within a few hours.

The Saharan snow isn't the only example of desert snow. In early 2011, freak snowfalls in the Andes covered the Atacama Desert in Northern Chile. The typical rainfall in this area is less than three millimeters per year (that's twelve one-hundredths of an inch).

On January 11, 2011, snow was on the ground in forty-nine out of fifty states in the United States. While you might guess Hawaii would be the snowless one, two of its peaks—Mauna Loa and Mauna Kea—regularly receive snowfall. The only snowless state on this date was Florida. Leading up to this record-breaking event, two storms covered the mid-section of the U.S. from Texas to the northern plains, while another pushed up the Gulf Coast and East Coast. At the same time, a nor'easter covered the Northeast. High peaks in the West (which are usually covered in snow all winter) accounted for the rest of the snow.

Residents of Baghdad, Iraq, got their first snow in one hundred years in 2008. In August 2011, Auckland, New Zealand, experienced snow for the first time in eighty years. Buenos Aires, Argentina, got its first snowfall in nearly a century in 2011. Most people in the city had never seen snow in their lifetimes. (The last recorded snow was in June of 1918.) A blast of polar air from Antarctica combined with extreme humidity to create sleet, which later turned to snow. The storm lasted ten hours.

SNOW DONUTS

Call it a perfect storm. If the temperature is hovering around freezing, the winds are blowing, and the snow is icy and crusty so that newly fallen snow won't stick to it (but rather bounces along it), you may be able to order up some snow donuts. The donuts (or *snow rollers* as some call them) form in the above conditions where there are irregular drifts sticking up from the snow or there's a steep surface, such as a cliff. When the wind blows a chunk of the icy snow, that chunk acts as a seed, so to speak, for the donut. The wind spins the chunk around, and layers of snow form around it.

Because of what meteorologists call "the pinwheel effect," the center develops a hole. The spinning snow forms into something resembling a white Life Saver, which gets larger as snow rolls downhill. Some snow donuts become long and tube-like (still with the empty center). The U.S. National Weather Service calls snow donuts a very unusual phenomenon. One of the largest snow donuts ever recorded was found in the state of Washington, measuring approximately twenty-six inches tall with an eight-inch-diameter hole.

SNOW CONES

They say that no two snowflakes are exactly alike, but they do generally tend to be hexagonal and symmetrical. However, on rare occasions, conical snowflakes have been spotted.

In 1931, a scientist at the University of Michigan reported conical snow falling on three occasions. The snow was described as resembling "the conical section of a sphere." The odd-shaped flakes were thought to be the result of many thousands of snowflakes densely packed together. The only other recorded instance of conical snowflakes comes from another scientist working at a weather station in Utah in 1920.

THEY MIGHT BE GIANTS

In 1951, postmen doing their morning rounds in the English city of Berkhamsted observed that giant snowflakes "as big as saucers," were falling from the sky. The snowflakes were said to measure five inches in diameter. Those are dwarfed in size, however, by the snowflakes "as big as milk pans" that were reported in Montana in 1887, which were said to measure fifteen inches across. German meteorologists were able to record snowflakes of four inches in diameter during a 1915 storm. The flakes were also of a strange shape, said to be like "bowls with upturned rims."

Giant snowflakes are the result of many snowflakes sticking together. The more complicated the structure of a snowflake's "arms," the more likely it is that it can adhere to another snowflake. There's speculation that this phenomenon also takes place when the temperature is just above freezing, making the flakes wetter and stickier. Altitude also plays a part. Flakes that start to bond when higher up in the atmosphere have more time to grow as they're falling.

Snowflakes "as large as basketballs" may fall daily in Arctic areas, scientists say, in extremely remote areas of North America and Asia. By flying into snowstorms, they have been able to record snowflakes two to three inches in diameter, but they note that flakes much larger than that may exist in the Arctic. Since these areas are uninhabited, it's difficult to conduct research and collect data.

SUMMER SNOW

White Christmases aren't a familiar sight in Australia. Christmas takes place during the southern hemisphere's summer, when temperatures can reach into the 80s (Fahrenheit) or even higher in desert areas. But in 2010, residents of New South Wales and Victoria kept their fingers crossed for a white Christmas when they got four inches of snow in late December. Meteorologists reported that some areas of Australia received eleven inches of summer snow, along with the lowest temperatures seen in five decades. Granted, the areas that received snowfall were in the higher elevations, including Mount Hotham and Mount Buller in Victoria. Some towns in the area received their first snow in recorded history during this period.

RAINBOW SNOW

On January 31, 2007, residents of Omsk, Russia, woke up to snow outside their windows. That wasn't so strange—snow covers the ground in this Siberian city from January to April. But this snow didn't look, feel, or smell like regular snow: it was pale orange, oily to the touch, and smelled—depending on whom you asked—like rotten vegetables or dirty gym socks.

Some witnesses said it looked as though the world had been dipped in orange sherbet. Others said the drifts of orange looked like the surface of Mars. Warnings went out from TV and radio stations: Please don't eat the orange snow, and keep your pets and farm animals from eating it, too.

Russians have good reason to worry when strange things fall from the sky: in 1986, an explosion in Chernobyl, Ukraine (then part of the Russian Soviet Union), sent deadly radioactive material drifting thousands of miles across the sky, poisoning the air and water, and causing major health problems for people in neighboring countries. So when orange snow fell in Omsk, suspicions that it was toxic weren't altogether unfounded. The Russian government and even the international environmental group Greenpeace sent a team of scientists to test samples of the snow. They soon found out that, although it wasn't fragrant, the snow wasn't harmful to humans or animals.

So what caused the stinky, sherbet-like snow? Scientists found that the iron content in the snow was four times the normal level. Iron is known for its red color, and is found in sand, clay, soil, and dust particles, all of which were found in the snow in Omsk. When the snowflakes formed, it's possible that they attached themselves to reddish-colored dust or soil.

As the investigation continued, all signs pointed to a dust storm in Kazakhstan the week before. The strong winds, scientists said, kicked up dust particles from close to the surface of the earth, sweeping them into the troposphere (where weather is made). At that high altitude, the particles might have mixed with the ice crystals that eventually turned into snow. The day the snow fell, the wind had been coming from the southwest, which was in the direction of Kazakhstan. As far as investigators were concerned, the case was closed. But what about the smell and the oily feel? No official explanation has yet been offered.

Meanwhile, purple snow fell on the Russian city of Stravpol in 2010, but the cause was not immediately known. Scientists eventually pointed to dust blown in from Africa, but environmentalists chalked it up to pollution. In 2011, gray and beige snow fell on the Russian city of Samara. This snow was also blamed on a Kazak dust storm.

SNOW-HOW

To understand weird snow, it helps to understand how regular snow is made. It's all part of the water cycle. Water from the earth evaporates into the air, rises up, and condenses into clouds. When the clouds get too full, precipitation falls to the earth and accumulates, starting the cycle all over again. Snow happens when certain conditions are in place during the cycle. The vapor droplets in the clouds have to have something to attach themselves to—usually a tiny dust particle—and the temperature in the clouds has to be cold enough for the droplets to condense directly into ice crystals rather than into rain or sleet. Once the crystal forms, more vapor can attach to it, creating a snowflake. There's one point in the process when strange things can happen. If that little dust particle happens to be a very strong shade of red or yellow, the snowflake might be, too.

GREAT MOMENTS IN STRANGE SNOW HISTORY

"Blood snow," which has a distinct red color, has been reported in Europe since Roman times. In 1755, six feet of blood snow fell in the

Alps. These snows are usually explained as the result of sandstorms in the Sahara sweeping particles into the atmosphere and redepositing them in the snow.

"Bug snow" is more common than you might think. In 1749, people in a certain area of Sweden reported that their hats had live bugs and worms clinging to them after a snowfall. In Switzerland in 1922, caterpillars, spiders, and ants were said to have fallen from the sky along with snow.

In January 1887, snowflakes as long as a men's size-thirteen shoe and as thick as a mattress fell in Montana. Lots of snowflakes clumping together to form larger ones was cited as the cause.

Yellow snow fell in Bethlehem, Pennsylvania, in March 1869. Scientists chalked it up to the presence of pollen in the air, blown up from the southern states by heavy winds.

In 1895, residents of Alma, Colorado, found themselves in the middle of a pink snowstorm. When the storm was over, those who'd been caught in the snow reported muddy patches all over their clothes.

WATERMELON SNOW

July is the month for watermelon—whether it's the plump, juicy watermelon growing in the garden, or pink, fragrant watermelon snow ripening across the highest mountain peaks. When the sun finally reaches into the places where drifts run deep, microscopic red algae start to bloom under the snow. The algae paint the slopes with streaks of red and pink and emit a surprisingly watermelon-like scent.

Few plants or animals can live in the harsh environment above ten thousand feet, but red algae, also called snow algae, thrive there. The algae even support a whole mini-ecosystem of creatures that feed off it, such as snow worms, snow fleas, and water bears, which are tiny creatures with claws on the ends of their legs. The algae's red color—caused by *carotenoid*, the same pigment found in lobsters, carrots, and tomatoes—protects it from the strong ultraviolet light found at high elevations, but it also tints the snow around it. Hikers report that the color is so strong that it stains the bottoms of their boots and pants.

Watermelon snow has been found wherever there is snow: in the mountains of Asia, Africa, Australia, and even in seemingly tropical New Guinea. Early Arctic explorers were surprised how white Greenland was until they found something even more unusual, the so-named Crimson Cliffs along its shore. At the time they thought the red color might have come from iron deposits from meteors, but now scientists know it came from a whole lot of snow algae.

Snow algae don't just come in red—green, blue, and orange snow algae grow at lower elevations. No matter how tempting it might be to taste it, avoid doing so since the bacteria can cause painful stomach problems.

THUNDERSNOW

While snow typically falls to the ground in peace and quiet, the unusual phenomenon of thundersnow turns that image on its head. Thundersnow is just what its name implies—thunder and lightning during a snowstorm. It forms when warm, moist air from the area just above the ground makes contact with colder air higher up in the atmosphere (just like a regular thunderstorm). Thundersnow comes with the typical thunder rumbles, but instead of bolts of lightning, observers on the ground usually just see flashes of light from behind the clouds or a uniform brightness in the clouds.

Thundersnow is experienced most commonly in the late winter or early spring. While it can happen anywhere, thundersnow has been known to happen more frequently in the Great Lakes or the Colorado Rockies. With thundersnow comes heavier downfalls of snow, as much as two to four inches per hour.

All Hail!

Hailstorms are fairly common. They're formed when there's an updraft into a mid-altitude-level storm cloud in which the temperature is below freezing. They're usually associated with thunderstorms and can also be linked to tornadoes. Hail usually occurs inland rather than in coastal regions.

In 2011, a scientist at Montana State University made a groundbreaking discovery in the study of hail. At the center of hundreds of hailstones he and his research assistants analyzed, they found microscopic germs, suggesting that the microbes serve as water magnets that attract droplets, which form ice crystals and then, hail. That could mean that microbes are the real cause of hail, not dirt particles, as scientists had previously believed.

WHERE THE HAIL IS IT?

Kericho, Kenya, is said to be the most hail-prone place in the world, with up to 50 days of hail each year. In 1965, it hailed on 113 separate days.

Northern India and Bangladesh are known for producing the largest hailstones and deadliest storms.

In North America, the area where Colorado, Nebraska, and Wyoming intersect is known as Hail Alley. Seven to nine hailstorms occur there each year between March and October.

Amarillo, Texas, and Wichita, Kansas, are the most hail-prone cities in the U.S.

In Europe, Germany is the sight of the most numerous hailstorms. A 1984 hailstorm in Munich caused $2 billion worth of damage.

ALL HAIL BREAKS LOOSE

The supercell hailstorm of Sydney in 1999 was the most expensive natural disaster in Australian history. Half a million tons of hail fell on the city during the course of the storm. The stones fell at a speed of two hundred miles an hour.

The most expensive hailstorm in U.S. history happened in St. Louis in 2001, when one- to three-inch hailstones rained down on the city. The outcome was over $2 billion in damage.

The city of Huancavelica, Peru, is said to have endured the longest hailstorm in history. For more than three hours in one day in 2010, the area was pelted with hailstones.

In 2007, a nearly three-hour hailstorm left residents of Bogotá, Colombia, with cars buried in ice. Many sustained injuries and had to be treated for hypothermia because they were wearing light summer clothing at the time of the storm and were caught for hours afterward in the drifts of hail.

SUPERSIZE IT

A normal-sized hailstone is roughly the size of a dime, although sometimes more like a penny or even a marble or grape. But then there are the exceptional hailstones. You've probably heard of some compared to the following things: golf balls, oranges, goose eggs, baseballs, tennis balls, grapefruit, and even bowling balls. That's when you start to get into the territory of *megacryometeors*, or supersized hailstones. These hefty hailstones form when multiple hailstones amass into one on their way down to the ground (some even form into mega-stones after they hit the ground). A megacryometeor is classified as any hailstone of unusual weight. More than fifty megacryometeors have been recorded worldwide since the year 2000.

Supercell thunderstorms frequently produce what storm chasers call "gorilla hail." There's no official size limit on these hailstones, which are defined as abnormally large and dense stones that fall at tremendous velocity.

MEGA-HAILSTONE HISTORY

1822: Bangalore, India. Hailstones the size of pumpkins were reported to have fallen, killing local cattle on impact.

1826: Khandesh, India. A hailstone one-cubic-yard long was reported.

1838: Dharwar, India. Locals reported a hailstone "the size of an elephant," which measured twenty-five feet in diameter and weighed hundreds of pounds.

1849: Isle of Skye, Scotland. A twenty-foot-diameter hailstone was reported.

1882: Salina, Kansas. Railroad workers reported being pelted with four- to five-pound hailstones. Later the same day, an eighty-pound hailstone was found on the ground, but researchers say it only got that big after it hit the ground.

2010: Vivian, South Dakota. The largest hailstone ever recorded in the U.S. was found after a severe thunderstorm. Measuring eight inches in diameter, eighteen inches in circumference, and weighing nearly two pounds, the hailstone was described as "just short of the size of a soccer ball." The man who retrieved and stored the stone said that it had actually melted a little in his freezer so it was initially larger.

DEADLY HAIL

A single blow from a falling hailstone can be enough to kill. In Moradabad, India, in 1888, 246 people were either pounded to death by hailstones the size of "goose eggs" or had died from exposure after being buried in piles of the hailstones for days.

In Russia in 1920, twenty-three people were killed by hailstones weighing one to two pounds each.

In China, twenty-five were killed in a hailstorm in 2002 and eighteen people were killed by hailstones in 2005. Two hundred were reportedly killed by hail in 1932, but the claim can't be verified.

Only three deaths in the U.S. have ever been attributed to hail, the last one being in 1979.

THE SHAPE IT'S IN

Common hailstones are round, of course. But many people have been surprised to be pelted with hailstones with a more sculptural quality.

German records from 1393 report hail shaped like "human faces with spiky beards."

In 1853 in Peshawar, India, eighty-four people and three thousand oxen were killed by a hailstorm that rained hailstones as large as one foot in diameter. The stones were said to be spherical in shape.

"Tadpole-shaped" hail was reported in Bavaria in 1881. It supposedly had a kind of tail on the end of an oblong shape, with five, equidistant "knobs" on each side.

In 1888, scientists in England observed disk- and "Saturn-shaped" hailstones. They were round in the center, but also had protrusions around the center that resembled the rings around the planet Saturn.

Flat, "sheet-like" pieces of hail fell in Oregon in 1894.

In 1923, "pyramidal" hailstones are said to have fallen in Aberdeen, Scotland.

In 1948, residents of Hot Springs, Arkansas, observed donut-shaped hailstones. In the same year, star-shaped hail was reported in Kenya.

While working on a roof in Germany in 1951, a man was hit and killed by a six-foot-long "spear" of ice.

In 1968, scientists in Oak Ridge, Tennessee, reported hailstones that were shaped like stars and daggers or had spike-like protrusions.

"Cubic" hailstones have been reported in various parts of the world on numerous occasions.

DEEP, WIDE, AND STRONG

Unlike snow, hail doesn't form a compact layer when it hits the ground, which means deep layers of hail can accumulate quickly during a heavy storm. With a one-inch accumulation, the ground will appear white on top of grass. With two inches of hail, it will look like the aftermath of a winter snowstorm.

In 1823, a hailstone in Hyderabad, India, produced "enough hailstones to cool the wine for several days."

A French hailstorm in 1865 yielded a five-yard-deep layer of hailstones.

A hailstorm in Pasadena, California, in 1884 left two inches of hail on the ground. Since the storm brought everything to a grinding halt, people celebrated the day as a "hail holiday."

In 1959 in Seldon, Kansas, over eighteen inches of hail accumulated during a single storm, covering an area fifty-four miles wide.

In Clayton, New Mexico, in 2004, fifteen-foot-tall "hail cliffs" formed when a heavy rain followed a heavy hailstorm. Water pressure from the swelling creeks pushed the hail into the creek beds to make the cliffs.

In 2008, in Ottery St. Mary in Devon, England, a two-hour hailstorm resulted in twelve inches of hail, which accumulated into six-foot drifts. The town was encased in "a river of ice" according to newspaper reports. The accumulation measured 270 million cubic feet of hail.

ICY PLUNGES

Since the beginning of the commercial airline industry, there have been frequent reports of ice falling from the sky. This can happen when buildup from the wheel wells or vertical stabilizers breaks loose and drops down through the atmosphere (this tends to happen as the plane is preparing to land). There's also airline lavatory waste, a mixture of human waste and disinfectants, which leaks from an airplane, freezes in the high altitudes, and falls to earth. This offload is often called "blue ice" and can be identified by its color, which is from the disinfectant mixture. However, there are numerous incidents of blocks of ice falling from the sky, which have nothing to do with airplanes. That kind of ice is called "pure ice" or "cloud ice."

1953: New York, New York. A fifty-pound ice ball was found on the ground near LaGuardia Airport, although no one could actually claim to have seen it fall. Those who observed it said it was pure ice and not blue ice, despite the proximity to the airport.

1973: Manchester, England. A noted physicist was almost hit by a twenty-one-ounce piece of pure ice that fell near him. He preserved it in his freezer.

2007: California. A "bowling ball-sized" piece of ice broke through a roof, hitting a boy on the head. In the same year, a fifty-pound block of ice broke through the roof of an Iowa home.

2008: Oakland, California. A block of ice fell from the sky into Bushrod Park, leaving an impact crater three feet deep. Observers say that there was a streak of light, a whooshing sound, and a big bang, after which the grass-covered field was covered with shattered pieces of ice the size of human heads. When firefighters retrieved the ice from the crater, they found an intact piece that was approximately three by three feet. Local meteorologists were at a loss to explain the cause of the ice. "Sometimes big balls of ice just drop from the sky without any real explanation," one was quoted as saying.

2009: Pennsylvania. A massive piece of ice broke through the roof of a home and shattered into three pieces, one weighing more than six pounds.

FREE GIFT INCLUDED

Frogs, fish, and worms fall from the sky: this much we already know. But sometimes they come gift-wrapped inside hailstones. One of the strangest hailstone stories comes from Bovina, Mississippi, where, in 1894, a live six-by-nine-inch turtle was found encased in a hailstone. The same storm brought a block of alabaster inside a hailstone.

Seventeenth-century records from England describe the presence of wheat and seeds encased inside hailstones.

Worms were found encased in hailstones during a storm in 1713 in England.

The "blood hailstones" that fell in 1873 in Italy turned out to be filled with dirt.

An 1882 hailstorm in Iowa brought two live frogs encased in hailstones.

Blocks of ice with carp frozen inside them fell in Essen, Germany, in 1896.

HAIL STRIPES

Through the years, farmers have discovered the presence of "hail stripes" in the damage done by hailstorms. They're created when one strip of crops is ruined by hail while an adjacent strip remains unharmed. The stripes can vary in length and width.

FLAVORED ICE

It's only natural for kids (and sometimes grownups) to taste snow, so it's not surprising that some have also sampled hailstones. What's surprising is that some hailstones actually appeared to be flavored. In 1871, turpentine-flavored hail was reported in Mississippi. In 1874, hailstones flavored with carbonate soda (the kind of soda used for carbonated beverages) were reported in New Jersey. Sugar-flavored hailstones were reported in India in 1893.

SONIC HAIL

When hail hit the pavement during a 1911 hailstorm in Missouri, observers reported hearing "sharp reports" similar to the sound of a pistol shot. The hailstones splintered upon impact and bounced around, leading those watching to describe the scene as reminiscent of popcorn popping. Meanwhile, during an electric storm in 1874 near Pike's Peak in Colorado, there were reports of a crackling sound as hail hit the ground.

While these incidents point to hail making noise, through the centuries people have resorted to using noise to prevent hail. Hail cannons were invented in the eighteenth century as a measure to stop hail, which frequently causes crop devastation. The device emits sound waves that are said to disrupt hail formation. Some say that the sound is like that of a real cannon, a deep boom that can reverberate for hours.

While scientists cannot explain the theory behind these devices, some farmers say they really work. Modern hail cannons came back into use in the twenty-first century. They can be activated whenever a potential hailstorm is spotted on the radar. The booms are emitted every six seconds for up to thirty minutes. In at least one case, modern hail cannons have led to lawsuits for disturbing the peace from those who live in the vicinity.

LAZY HAIL

A hailstone's rate of descent depends on its size and density. Typically, the heavier it is, the faster it will fall. But there have been reports of hail that falls unusually slowly for its size. During an Iraqi hailstorm in 1930, slow hail was reported falling at only nine miles per hour—when the typical rate for hailstones of equivalent size would have been thirty miles per hour. This could be explained by wind shear that slows the hail down rather than speeding it up.

It Came from Outer Space

Sometimes the things that fall from the sky don't come from our atmosphere at all. Or, at least we don't think they do.

WOULD YOU LIKE PEANUT BUTTER WITH THAT?

Star jelly is one of the nicer names given to a strange gelatinous substance that has been found on the ground in isolated incidents throughout history. It's also been called "star powder," "star spawn," "star slough," and, perhaps the most unappealing nickname, "star rot." While there are many variations of star jelly found in different places, in general, it's described as a translucent substance that looks something like a jellyfish, but without the same shape. In fact, it's been called blob-like, and star jelly found in Philadelphia in 1950 inspired the horror film *The Blob*, about a gooey monster that comes from outer space. In reality, teens found "a domed disk of quivering jelly, six feet in diameter, one foot thick at the center, and an inch or two near the edge," and called the police. When one officer was brave enough to touch it, it dissolved into a "scum." In the UFO-crazed atmosphere of the 1950s, the incident made headlines across the U.S. It was quickly linked to a meteor sighting that had been reported the night before.

But star jelly reports go much further back in history. A thirteenth-century scholar wrote about *stella terrae*, Latin for "star of the earth" or "earth-star," and described it as "a certain mucilaginous substance lying upon the earth." Fourteenth-century physicians called it *uligo* or *sterre slime*, "a certain fatty substance emitted from the earth, that is commonly called 'a star which has fallen.'" In 1656, the metaphysical poet Henry More observed that "the Starres eat those falling Starres, as some call them, which are found on the earth in the form of a trembling gelly, are their excrement." In Veracruz, Mexico, star jelly was historically referred to as *caca de luna* or the Moon's excrement.

The known facts about star jelly are that it is only found in the morning and evaporates when temperatures rise or when it's touched. It's alternately described as foul smelling and odorless and ranges in color from translucent to white to gray. There are some reports of goo with a greenish or purplish color and some have described it as "luminous."

Scientists doubt the association of star jelly with meteors, insisting that such a substance would be incinerated in the upper atmosphere when a meteor enters the atmosphere. The other most common explanation for star jelly is that it is animal vomit—specifically frog spawn regurgitated by birds or other animals that might be prone to eat such things. The theory is that the frogs eggs are swallowed before ovulation, making it "regurgitated egg duct," which swells when wet. The flaw with this theory is that there has never been any amphibian DNA identified in the jelly. Another theory is that the jelly is *nostoc* or slime mold, a type of blue-green algae that can swell up after rainfall, found in colors from white to pink, purple, orange, brown, and yellow. These molds are easily dissolved by wind or rain.

STAR JELLY JOURNAL

1846: Loweville, New York. A "body of fetid jelly, four feet in diameter" was found after a streak was seen in the sky "larger than the sun."

1855: Koblenz, Germany. After observing a falling star, witnesses found a grayish blob in a field. It "trembled" when poked with a stick.

1908: Pembrokeshire, England. A blob found in the grass was sent to a botanist for analysis and was determined to be "bacteria."

1909: Allegheny County, Pennsylvania. A blob with an "intolerably offensive smell" was found after the observation of a falling star.

1979: Texas. Purple "blobs" were found after the August Perseid meteor shower. Scientists determined that it was discharge from a local battery factory.

1983: North Reading, Massachusetts. A grayish-white, oily gelatin was found on lawns, the street, sidewalks, and on gas pumps.

1994: Oakville, Washington. People reported a "rain gelatin" in the region.

1996: Kempton, Australia. The morning after a meteor streaked through the sky, witnesses reported clear jelly found on the grass and sidewalks. It was described alternately as fish eggs or baby jellyfish.

2009: Scotland. Gelatinous blobs found in Scotland were photographed and analyzed, leading to no clear conclusion. Speculation included that it was seagull vomit, the stuff from inside diapers, and "alien poo."

2011: Alaska. A yellowish-orangeish "gelatinous slime" was found on the ground and in rain barrels in a remote Alaskan village. Because some of it was found on the roof, it was presumed to have fallen from the sky. The substance was analyzed and found to be a type of parasitic spore.

ROCK STARS

Meteorites entering Earth's atmosphere are a common occurrence. They do so every day, although they typically don't survive the journey through the upper atmosphere. While it's impossible to count all of them, estimates point to around five hundred per year, ranging in size from marbles to basketballs.

The largest meteor impact event on Earth in recent history, the Tunguska Event of 1908, created a tremor that measured the equivalent of an earthquake of 5.0 on the Richter scale (considered an earthquake of moderate size). Eighty million trees were snapped in half over an 830-square-mile area as the meteor, which was never found, sped into a remote, largely unpopulated area of Siberia.

Those who witnessed the event lived in nearby towns or small villages. They reported seeing a streak of blue light flashing across a cloudless sky, which headed downward for more than ten minutes. As the streak got closer to the ground, a cloud of black smoke arose and strange flames started to emerge from it. A huge boom that sounded like artillery fire shook the buildings in surrounding areas and hot winds arose that were strong enough to knock people off their feet. The ground began to tremble as if there were an earthquake, and then a rhythmic thumping sound was heard and a rattle came from under the ground. Windows were shattered up to hundreds of kilometers away, and iron locks were snapped.

The sky over Asia, and as far away as Europe, glowed for several nights. The atmospheric dust from the fall was present for months afterward. Scientists now speculate that the event was just an "airburst" from a comet or asteroid traveling up to six miles above Earth and that there was no actual meteor that struck the ground.

HAMMERSTONES IN HISTORY

Meteorites that hit buildings or beings are sometimes referred to as "hammerstones."

1911: Egypt. A dog is said to have been hit by the Nakhla meteorite —the only known fatality from a meteorite.

1938: Pennsylvania. Two small pieces of meteorite fell in a rural area, one hitting a cow.

1954: Sylacauga, Alabama. A woman was hit by a meteorite when it crashed through her roof.

1992: Peekskill, New York. A meteorite weighing more than twenty-six pounds crashed into a 1980 Chevy Malibu.

2003: New Orleans, Louisiana. A forty-four-pound meteorite crashed through the roof of a home, and then down through two stories, until it exited through the floor and became embedded in the ground.

2003: India. Several people were injured when a spray of burning meteorites fell on a village, setting fire to several homes.

2008: Sudan. A rain of more than six hundred meteorites was identified by scientists as parts of an asteroid collision. It was the first time scientists were able to track an asteroid through space, as it entered the earth's atmosphere, and then all the way to its point of impact.

2010: Virginia. A meteorite crashed through the roof of a dental office in Virginia. It was described as being composed of chondrite and weighing eight grams (0.035 ounces).

2010: Bosnia. After his home was hit by meteorites for the sixth time in three years, a Bosnian man proclaimed that his home was being targeted by aliens.

2011: Poland. A stone weighing more than two pounds hit the roof of a building and broke into pieces.

2011: Kenya. Residents of Kilimambogo and Tala described hearing a sound as loud as a bomb or an airplane crash and then finding a smooth stone that had dug a hole in the ground. The rock was black and weighed about eleven pounds. Witnesses said the rock raised a cloud of dust as it hit the ground and spun around on impact. It was hot to the touch.

SPACE DUST

Until recently, meteorologists didn't believe that the fall of meteorites had any effect on the weather. It was thought that the particles they brought into our atmosphere were too tiny. But new research indicates otherwise. After a meteor fell in Antarctica, it left more than a metric ton of dust in the atmosphere, creating a cloud over the fall area.

Inspecting the dust, researchers found that the particles were big enough to form the basis for raindrops. While that still wouldn't make much difference in the case of a small meteorite, a large one could cause "weather disturbances," such as rain and possibly thunderstorms. If a meteorite fell near the equator, the debris could be strongly affected by solar radiation. The researchers also found that the impact of the meteor had an effect on the ozone layer, which was previously thought to be strong enough to withstand any sort of penetration by meteors. Findings suggest that it is not as strong as previously believed.

Any Way the Wind Blows

We depend on the wind for energy, cooling, and even for transportation. But it's anything but dependable. It spins and swirls out of control with unbelievable force. It gusts and thrusts at mind-boggling speeds, destroying everything in its path. It can be playful, kicking up dust in its wake, or perilous, burying cities in sand. When wind and water get together, you can also expect some surprising results. In this chapter, we'll look at many of the wicked ways the wind blows, bursts, and bowls us over.

Devilish Dust

When hot air near the ground rises quickly and hits a pocket of cool, low-pressure air, it can start to spin, forming a column of air. As that column stretches out, the spinning motion increases, and that spinning heats the air even further, creating a vortex. More hot air enters from the bottom of the vortex and the cooler air on the outside acts as a counterbalance, keeping the spinning column stable. Once it has stabilized, the spinning (along with the friction from the surface) creates a momentum that allows the funnel to move. As it moves, it picks up dust along the way. That's a dust devil.

DUST DEVIL FACTS

Dust devils aren't associated with storms. They tend to form over flat areas where there's little vegetation and high temperatures close to the surface. Deserts, where the temperature difference between surface and air is extreme, are perfect for dust devils. The presence of wind will destabilize the funnel, causing the dust devil to dissipate.

Dust devils can range in size from tiny (just one-and-a-half-feet wide and a few yards tall) to more than thirty-three feet wide and one thousand feet tall. They rarely reach speeds over forty-five miles an hour and typically last no more than a few minutes. Extreme dust devils can reach speeds of over sixty miles per hour and last up to twenty minutes. A dust devil can lift .035 ounce of dust per second from each square yard of ground it moves over. That means a big dust devil could lift seventeen tons of dust in thirty minutes. Because of the friction caused on the surface by the whirling wind and debris, sparks can sometimes be seen coming out of dirt devils, which is often mistaken for lightning.

A DUST DEVIL TIMELINE

Despite their threatening name, dust devils typically don't cause too much damage. But rare, powerful dust devils can cause injury and even death. In 2000, a dust devil that ran through an Arizona fairground is said to have reached speeds of seventy-five miles an hour, which is the speed of the lowest level of tornado.

A man was killed in Maine in 2003 when a dust devil lifted the roof off a building when he was inside.

A woman was killed in a shed in Wyoming in 2008 when a dust devil caused the roof of the shed to collapse.

In 2009 in Makaha, Hawaii, teachers at Makaha Elementary School reported that a double dust devil event picked several fourth graders off the ground and smashed them into a fence. According to those on the scene, the day was clear when they saw a pair of dust devils racing toward a class that was outdoors for PE class. The dust devils lifted items from the field and spun them, including picking up kids and sending others tumbling down a hill. Several children were hit with flying tennis rackets.

In 2011, an inflatable bounce house was lifted up by a dust devil in Tucson, Arizona, just after parents got their children out of it. It was the third Arizona incident involving bounce houses and dust devils in that year alone.

DEVILISH LORE

In many areas of the Arabian Peninsula, dust devils have long been associated with the presence of a *djinn* or genie, a mischievous spirit who lives inside the column of dust. In Brazil, the prankster Saci is said to live inside a dust devil. According to legend, he grants wishes to anyone who can steal his magic cap.

In 2010, a rare dust devil was reported in Ireland by a man who noticed straw being tossed high in the air over a nearby field. Meteorologists said that although mini-tornadoes do happen on occasion in the country, dust devils are very unusual. In the old days, such occurrences were said to be the work of the fairies.

DUST DEVIL NICKNAMES

Dancing devil

Dirt devil

Dust whirl

Sand auger

Sand pillar

Chiindii: A Navajo term for ghost or spirit; good spirits spin clockwise, bad spirits spin counterclockwise.

Will willy, or whirly whirly: In Australia, aboriginals believed these storms to be malevolent spirits that would chase or abduct naughty children.

Ngoma cia: (a.k.a. a Kikuyu): A Kenyan word for woman's demon.

Redemoinho: From Brazil.

Remoinho: From Portugal.

Fasset el 'afreet: From Egypt, meaning ghost's wind.

FOREIGN DEVILS

A cousin of the dirt devil, the *haboob* is a kind of dirt devil found in desert areas such as the Sahara, the Arabian Peninsula, Central Australia, Arizona, Mexico, and Texas. Unlike dirt devils, haboobs form during thunderstorms. When the storm releases precipitation, there's a downburst that sends sand blowing up, creating a wall of sediment that precedes the storm. Haboobs can send dust several miles up into the air and spread it up to sixty miles wide.

During a 2011 haboob in Arizona, some local residents complained to TV stations about the use of "a Middle Eastern word" to describe an "American dust storm."

RED PLANET, RED DUST

Dust devils aren't limited to Earth. NASA scientists have found that they're pretty common on Mars, too. Going back to the 1970s, Viking orbiters have snapped images of dust devils in action on the surface of Mars. Since rovers have started landing on the planet, there's even more photographic evidence. Although the atmosphere on Mars varies greatly from Earth's, dust devils form there in essentially the same way: with warm air just above the surface rising up and cool air above it falling, creating a *convection cell*. That cell will begin to spin if a horizontal gust hits it.

Martian dust devils can grow up to ten times as high and fifty times as wide as ones on Earth, meaning they could go as far as six miles high and three hundred miles wide! Scientists also report that Martian dust devils are electrified, with "mini-lightning" flashing out of them. Also, while traces of dust devils are usually erased from desert sands in a day or two on Earth, on Mars, dust devil trails may remain visible for weeks. Scientists say that dust devils aid their Martian research because they blow the dust off the rovers' solar panels!

GUSTERS

A gustnado is described by the U.S. National Weather Service as "a short-lived, ground-based, shallow vortex that develops along a front associated with either thunderstorms or showers." Although the wind in a gustnado may reach low-level tornado strength, gustnadoes are more closely related to dust devils than to tornadoes and don't have contact with thunderclouds. Gustnadoes are so weak and short-lived that most go unnoticed or unreported. In fact, they're most often recognized when they're moving over open ground because the sight of dust kicking up may be the only evidence of one. Gustnadoes are rarely strong enough to form a funnel cloud. Gustnadoes are most common in the Great Plains and Midwest area of the U.S., areas that are also prone to tornadoes.

In 2011, a gustnado was blamed for a violent blast of wind that knocked over a stage at an outdoor concert in Indiana, killing five people and injuring dozens of others. Meteorologists estimated that the gust that blew across the stage clocked in at fifty miles per hour.

LANDSPOUTS

A cousin of the gustnado is the landspout. Like tornadoes, landspouts form under cumulus clouds and reach up into the cloud base. They're stronger than gustnadoes and last longer, but they're not as strong as tornadoes. Meteorologists differ on the subject of landspouts, with some saying they're just tornadoes.

A Line in the Sand

Unlike dust devils, which form vortices, sandstorms (also called dust storms) are formed from strong winds without the spinning motion. Epic sandstorms are common in some areas of the world, such as the Sahara and Gobi Deserts, the Arabian Peninsula, Iraq, and even the southwest United States. During such storms, entire dunes can be picked up and moved by the wind.

SANDSTORM TIMELINE

In an unexpected sandstorm, a Danish seaside village was buried in sand up to the tip of its church steeple in 1775.

In an 1895 sandstorm in Colorado, nearly a quarter of the area's cattle population died from suffocation as they breathed in airborne dust.

Each spring, Egypt is hit with the *khamaseen*, a fifty-day wind that coats everything in dirt. A 1997 sandstorm in Egypt was so strong that it destroyed buildings and killed twelve people. The storm, which carried winds of sixty miles per hour, is said to have turned the sky first white, and then as dark as nighttime.

Sandstorms are also common in Saudi Arabia in the spring. A 2009 storm shut down the city of Riyadh and brought oil exports to a halt. Also in 2009, a week-long sandstorm that covered Iraq was described as the largest sandstorm in modern history.

A storm bringing sand from Africa buried Athens in yellow sand and fog in 2010.

A 2010 sandstorm in a German Baltic seacoast town caused an eighty-car pileup in which eight people were killed. With visibility suddenly at zero, the cars on the expressway crashed into each other at high speeds, many of them bursting into flames.

VANISHED

In 2009, what's believed to be the remains of a Persian army that "vanished" in 525 BCE was found in Egypt. The Greek historian Herodotus had written about an army that traveled to an oasis in the desert and disappeared around the same time a strong wind came from the south. The disappearance of more than fifty thousand soldiers became an enduring mystery. Italian archeologists working in Egypt came across skeletons, swords, and jewelry from the period that the soldiers disappeared. They were behind a large rock formation that could have served as a shelter during a sandstorm. The researchers speculated that the army got caught in a khamaseen and was buried under feet of sand.

DUSTED

Increasingly, sandstorms are being blamed on deforestation. In China, the sandstorm season lasts March through May. A 2002 sandstorm in Northern China affected more than one hundred million people. In Beijing, the city was covered with fog and the sky was "a weird yellow color." In 2010, officials in Beijing warned people to stay indoors and issued the highest-level pollution warning as dust blew through the city. The sky appeared orange, and statues in the city were covered with a layer of dirt. People wore surgical masks to go about their daily business.

Sandstorms are also common in the Australian outback. In 2009, a massive storm turned the sky "blood red." A 2011 storm turned the air in Sydney orange with sand blown in from the dry interior. It was called the worst dust storm in seventy years and caused the airport to close and thousands to report breathing problems.

WINDLESS

Meanwhile, as much as wind plays a part in weather, there are some places that are known for their absence of wind. The most notable area is the "horse latitudes," between thirty and thirty-five degrees north and south. There are different explanations for this nickname, but one that seems to have prevalence is that when ships coming from Europe to the New World reached this area, they were slowed by the lack of wind. The lengthening of the trip led to shorter supplies, and there were not enough provisions to feed horses; therefore, the horses were thrown overboard. This area is also known as the Calms of Capricorn and the Calms of Cancer.

The doldrums is another area of low wind close to the equator. In the past, ships tried to avoid being caught in the doldrums.

Spouting Off

Waterspouts are often described as "tornadoes over water." Rotating vortices, they form when cold air moves over warm water. The greater the difference between the air and the water temperature, the larger the waterspout will become. Waterspouts can rotate at speeds of more than sixty miles per hour (there's one report of a waterspout creating 200-mile-per-hour winds) and rarely last longer than fifteen minutes.

Whereas tornadoes connect funnels of air to the clouds, waterspouts suck water up into clouds. Waterspouts are fairly common in the summer months in the southern Atlantic, the Caribbean, the Adriatic Sea, and off the coast of Australia. They've also been seen over the Great Lakes and off the coast of Hawaii. The Florida Keys have more waterspouts than any other location in the U.S. There may be anywhere between four hundred and five hundred of them in any given season. Waterspouts are blamed for the many strange rains of fish, frogs, and other creatures described in pages 14 through 21.

SPOUT SLASHERS

In the past, sailors believed that demons lived inside waterspouts, and they would attempt to "cut" them with a black-handled sword, an action that was said to cause them to dissipate. As technology progressed, ships would shoot cannons into them in order to halt them.

DEADLY SPOUTS

Waterspouts typically don't cause too much damage, but there have been reports of boats getting caught up in them. In 1879, a waterspout caused the Tay Bridge in England to collapse just as a train was passing over it. Seventy-five people on the train were killed. A waterspout sank a shrimp boat in San Antonio Bay, Texas, in 1980, killing one person. A windsurfer was killed as the result of a waterspout on Lake Michigan in 1993. In a very strange incident in Alberta, Canada, a parasailer became disconnected from his parachute and towline during a waterspout and was blown a quarter of a mile away where he landed on a fence and was killed.

DOUBLE THE FUN

A boat in the South China Sea in 1967 reported witnessing a "forked waterspout," which came out of the cloud in a straight line and then divided into two branches well above the surface of the water. In another strange incident in 1928, a boat in the North Atlantic recorded seeing a waterspout that connected two clouds instead of connecting a cloud to the water. The ship's captain said that two waterspouts stretched between two cumulus clouds. There have also been reports of "double-walled" waterspouts that have a central core and another wall of water surrounding them. The waterspouts are said to be "pulsating." Meteorologists believe these double-walled waterspouts are optical illusions.

BLOWING OFF STEAM

Something of a hybrid between a dust devil and a waterspout, a steam devil is a kind of not-very-powerful whirlwind over water. In the same way that all whirlwinds form, steam devils form from warm steam rising from the surface of warm waters which hits cold air and creates a vortex. The vortex sucks in fog, thus making itself visible. Steam devils typically happen in the fall when lake water is still warm but the air in the atmosphere is starting to chill down. An ideal ratio for their formation would be 30°F air temperature to 60°F water temperature.

Steam devils were discovered in 1971 by a pair of meteorologists doing research at Lake Michigan. They're usually between one-hundred-and-fifty and six-hundred-and-fifty feet in diameter and can stretch up to one-hundred-and-fifty feet high. They're so slow moving that they typically don't get beyond a few rotations per minute and rarely last more than three or four minutes. (Some only exist for a few seconds.) Occasionally they will detach from their base and rotate across the lake or river and end up over land. However, that typically causes them to dissipate immediately.

Steam devils are most commonly seen on the Great Lakes but have also been reported at the southern Atlantic coast and in Hawaii, where they form from molten lava entering the ocean. They often form over geysers. If you want to see one, the geysers and hot springs in Yellowstone National Park are a good place to look because they occur there daily.

SNOW DEVILS

Snow devils (also known as snow whirlies) look similar to steam or dust devils but are caused by intense winds rather than the collision of hot and cold air. When a blast of wind hits an obstacle such as a hill, it can spin downward and lift up snow, creating a vortex. Snow devils rarely last more than a few minutes and usually don't get larger than thirty feet in diameter.

An Australian Antarctic expedition ran into a snow devil that was strong enough to lift 1,500-pound objects from the ground.

FIRE WHIRLS

Water and dust aren't the only substances that whirl. In the same way that those vortices form, a vortex can form in the presence of a wildfire. Most often called fire whirls, they're also known as fire tornadoes or fire devils. They look like rotating columns of fire rising up through the air (similar to the one evoked by Moses in *The Ten Commandments*). Because they develop inside widespread fires, they carry debris, ashes, and smoke up with them and spread the fire even further. Fed by the gas created by the fire, they can blow up and expand rapidly, rising to more than one hundred stories tall.

Fire whirls have also been reported in the presence of volcanoes. Observers of an Icelandic volcano saw fire whirls towering off the side of the volcano. They're also associated with earthquakes. During Japan's Great Kanto earthquake in 1923, a fire whirl is said to have lasted fifteen minutes and killed thirty-eight thousand people.

Blowin' in the Wind

Here are a few more winds that can be bizarre.

FROM TOP TO BOTTOM

Katabatic winds are essentially winds that move from higher elevations down slopes at speeds that can reach hurricane level. They can be found all over the world, each with a different name, such as the Bora in the Adriatic, the Mistral in France, the Santa Ana in southern California, the Tramontana in the area near the Alps, and the Oroshi in Japan. In Greenland these winds are called Piteraq. The katabatic winds of Antarctica are known to be particularly long lasting.

In Slovenia in 2010, a Bora wind was clocked at 124 miles per hour. The Santa Ana winds are frequently blamed for California forest fires. In 2008, 70-mile-per-hour winds spread fire through the San Fernando Valley just outside Los Angeles. Typically occurring in the months of October through December, the Santa Anas have been nicknamed the "Devil's Wind" because of all the damage they've caused, mostly by spreading fires during seasonal droughts.

BURSTING OUT ALL OVER

A microburst is a small but very powerful and dangerous wind front that develops suddenly, lasts no more than a few minutes, and affects a very small area (no more than two and a half miles). Similar storms that affect a wider area are called *macrobursts*. Winds can reach up to seventy-five miles an hour during a microburst. Microbursts come in two varieties: wet and dry. Wet ones come with precipitation while dry ones produce virga, which is rain that evaporates before it hits the ground.

BIG LEAGUE BURSTS

In 1890, a microburst on Lake Pepin in Wisconsin is said to have capsized the *Sea Wing*, killing ninety-eight people.

In 1895, a microburst capsized the *Pride of Baltimore* in the Caribbean, killing the captain and some of the crew.

One of the most notable microbursts in history took place during the Great Depression in the area known as the Dust Bowl, encompassing parts of Oklahoma and Texas. The region had been experiencing a drought that lasted throughout the first half of the 1930s. Topsoil was dry and eroding, and dust storms were common in the area. On April 14, 1935, a day that became known as Black Sunday, twenty different "black blizzards" or black dust storms hit the Great Plains. Meteorologists now consider the storm to be a macroburst. Rolling clouds of dust covered an area hundreds of miles wide, picking up and carrying more than three hundred thousand tons of dust. People were unable to see their own hands in front of their faces and reported that the sunny afternoon turned as dark as night. Winds reached between forty and sixty miles per hour.

In 2006, a microburst hit Lawrence, Kansas, damaging more than half the buildings on the University of Kansas campus.

In 2010, a wet macroburst hit Queens, New York, with winds of 125 miles per hour. Some accounts reported over three thousand trees downed by the storm. Around the same time, a microburst was blamed for the sinking of the tall ship *Concordia* off the coast of Brazil. All aboard were rescued. Just a few weeks later, a wet microburst in Chicago carried winds of 100 miles per hour.

Twist . . . and Shout

Tornadoes are some of the most deadly, damaging storms on the planet. They form within thunderstorms, especially supercells. Like other whirlwinds, tornadoes are created when there's a big temperature differential between pockets of air. There also needs to be moisture present in the air, which will vaporize, creating heat as the water condenses into clouds. The big difference between minor whirlwinds, such as dust devils, and major ones, such as tornadoes, is that tornadoes originate in the clouds and make contact with the ground. Whirlwinds that originate in the clouds but don't make contact with the ground are known as *funnel clouds*.

TORNADOES AROUND THE WORLD

The U.S. reports the greatest number of tornadoes of any country in the world with approximately 1,400 tornadoes per year. Within the U.S., the greatest number of tornadoes occur in an area known as Tornado Alley, which encompasses a wide area between the Appalachian Mountains in the East and the Rocky Mountains in the West. Texas has the greatest number of tornadoes, followed by Kansas and Oklahoma. In 2005, Alaskans spotted their state's first-ever tornado. Because the state is so big and so many areas are unpopulated, it's possible that there have been twisters there before that haven't been seen.

In Europe, tornadoes have been reported in the UK, Ireland, France, Italy, Germany, the Netherlands, Poland, Sweden, Spain, and Finland. The UK has the most tornadoes per year of any European country, with around thirty-three per year.

Tornados have been reported in China, India, and Bangladesh. Historical reports claim that a tornado in Canton, China, killed more than 10,000 people, but it's possible that the estimate included those who died due to subsequent flooding in the area. A deadly tornado in Bangladesh killed 160 people in 1998. Bangladesh was also the scene of the deadliest tornado in world history, which killed more than 1,400 people in 1989.

Australia also experiences tornadoes. In 1918, the "Brighton cyclone" struck Melbourne, Australia. The storm was actually two tornadoes with winds of nearly two-hundred miles per hour (an F3 rating). Two people were killed. The deadliest tornado in Australian history took place in Kin Kin, South Queensland, in 1971. Three people were killed. The strongest Australian tornado (F4) occurred in Bucca, Queensland, in 1992.

Tornadoes are rare in South America, but in 1973, fifty-four people were killed by a tornado that hit a small town north of Buenos Aires. A Brazilian tornado was reported in 1984.

MEASURING UP

The Fujita scale (named for Dr. Ted Fujita, who invented it) rates tornadoes from F0 to F5. The scale goes like this:

Wind Speed	Typical Damage
F0 40–72 mph	Broken tree branches, damage to signs and boards
F1 73–112 mph	Lifts houses off foundations; pushes cars off road
F2 113–157 mph	Roofs torn off, trees uprooted, mobile homes destroyed
F3 158–206 mph	Well-built houses and buildings destroyed
F4 207–260 mph	Major devastation
F5 261–318 mph	"Incredible" storm, complete destruction

A tornado beyond F5 would be "inconceivable."

Around 75 percent of tornadoes fall into the weakest category, but the one percent of tornadoes that reach the violent levels (F4 and F5) cause the most damage. A major tornado can stretch a mile wide, and last for more than an hour.

INVISIBLE TORNADOES?

Rain-wrapped tornadoes are said to be "invisible." They're difficult to detect because the tornado is "hidden" behind curtains of rain. Because of this, meteorologists can't send out warnings quickly enough. The devastating 2011 tornado in Joplin, Missouri, was described by some as a rain-wrapped tornado.

FOR THE RECORD BOOKS

The Tri-State Tornado of 1925 broke all kinds of records. The storm, which ripped through Missouri, southern Illinois, and southern Indiana, was the deadliest tornado in U.S. history, the longest-lasting, and the farthest travelling. It traveled a distance of 219 miles, lasting for longer than three hours and taking the lives of 695 people.

The St. Louis–East St. Louis Tornado of 1896 is thought to be the most damaging tornado in U.S. history.

April 2011 became the deadliest month for tornadoes with 116 deaths. In May of 2003, four hundred tornadoes were reported in the U.S., making that month the record holder for most tornadoes in a single month. The deadliest year for tornadoes was 1953 with 454 deaths from one thousand tornadoes.

SUPER OUTBREAKS OF TORNADOES

Between April 27 and 28, 2011, 178 tornadoes occurred in fourteen U.S. states, making it the biggest tornado outbreak in U.S. history. This outbreak broke the record previously held by the Super Outbreak of 1974, when 148 tornadoes swept through the U.S. and parts of Canada between April 3 and 4. The 1974 Super Outbreak remains the outbreak with the strongest storms: there were six F5 and twenty-four F4 tornadoes during this period. The 2011 outbreak caused 327 deaths.

UNSEASONABLE STORMS

Tornadoes are not uncommon in Missouri in the spring and early summer, but in January, they're practically unheard of. In January 2008, two tornadoes ripped through Missouri and Arkansas, killing two women. Meteorologists were stumped by the cause of these freak storms.

CAN A TORNADO REALLY DO THAT?

Thanks in part to *The Wizard of Oz*, tornadoes have been rumored to do strange things. Dropping houses on witches in a movie is one thing, but in the real world, can a tornado really pick up a house? Or pluck a chicken?

The chicken-plucking incident was recorded after an 1879 tornado, when a Texas tornado was said to have plucked thirty chickens. Modern researchers have concluded that while chickens may indeed be found featherless after a big twister, the condition is likely the result of "flight molt," which causes their feathers to fall out in times of intense stress.

Can tornadoes really carry people? There are numerous reports of it happening through the years. In 1955, a South Dakota girl and her pony were carried about one thousand feet through the air by a tornado. The girl landed on her stomach and held onto weeds to avoid being picked up again. In 1923, a baby is reported to have been picked up by a tornado and carried two miles away. She was found later in the day sleeping in a shed. A toddler was said to have been sucked out of a St. Louis bakery during a tornado and found unharmed three blocks away. In 2006, a man was sucked out of his Missouri trailer home and carried more than 1,300 feet, and then dropped—resulting in only a small head injury. Researchers say this is the longest recorded distance of a person being carried by a tornado and surviving. Still stranger are the reports of a man sucked out of his house and deposited in a tree, unharmed, in 1955, and a man pulled from his bed and found in a tree fast asleep.

There are numerous reports of people being carried aloft inside their cars or being sucked out of car windows. Some survive unharmed and others don't.

Animals are often caught up in storms, too. A 1915 tornado in Great Bend, Kansas, carried five horses from their barn and dropped them a quarter of a mile away, unharmed. In one nineteenth-century storm, a horse was carried up in a twister and deposited back on the ground with its saddle and harness still attached. During an 1899 tornado, three riders were reported to have been sucked into a tornado still on their horses, and then deposited on the ground again, all unharmed.

Cows inside twisters are fairly common. After a Saskatchewan storm in 1946, a live cow was found upside-down with its head imbedded in the ground. In 2008 in Arkansas, a cow was carried almost a mile and dropped down unharmed; however the same storm claimed the life of a man. During a tornado in 2010, a woman in Minnesota was hit by a cow that crashed through her home as it was carried by a tornado. The woman was pregnant at the time, but the baby was born with no complications. Her parents called her Skylar, in honor of her survival story.

During the 1915 Great Bend tornado, a tie rack is said to have been carried more than forty miles with neckties still attached to it.

After a Connecticut storm, a barn door was found ten miles away and a church steeple was found fifteen miles away. During a 1920 tornado, six jars of fruit were carried by a tornado and found uncracked. After a 1917 tornado in Kansas, the ticket window from a local train station was found intact in a nearby field. A cancelled check was found more than three hundred miles away after that same storm, which is said to be the longest recorded distance of debris flight from a tornado.

Tornadoes are often responsible for the freak effects of embedding items inside other items by the force of their winds. Perhaps the best-known of these stories is the bean that was found embedded in an egg in 1951. The egg was completely uncracked, even amidst the devastation of the storm. Pieces of corn, logs, knives, forks, and axes have been found embedded in trees. After an Australian tornado, a picture frame was found embedded in the wall of a room.

Perhaps the strangest story of all comes from a drive-in theater, which in 1996 was showing the movie *Twister.* You guessed it; it was struck by a tornado!

STRAIGHT-UP DEADLY

Strong enough to spawn several tornadoes, a *derecho*, Spanish for straight, is like a wall of wind followed by thunderstorms along a straight line (as opposed to the vortex of a tornado). They can stretch hundreds of miles and wreak massive havoc. Derechos feature winds of at least fifty-seven miles an hour, but they can reach eighty to one hundred miles an hour. They happen most often in spring and early summer in the Corn Belt of the U.S. (Iowa, Illinois, Indiana, Michigan, eastern Nebraska, eastern Kansas, southern Minnesota, and parts of Missouri) and along the southern plains, but they have been reported as far north as Michigan, as far south as Florida, as far west as Texas, and as far east as Maine.

Derechos form from what meteorologists refer to as *bow echoes*, which are thunderstorms that take on a curved shape. The first "super derecho" was recorded in 2009 when a storm ravaged Kansas with eighteen derechos clocking winds of seventy to ninety miles per hour. The storm was so strong that meteorologists compared it to a tropical cyclone, the kind of storm that creates hurricanes. The super derecho drove across one hundred miles, and crossed the Mississippi River into Missouri and Illinois before dissipating.

The Eye of the Storm

Hurricanes, cyclones, and typhoons are all types of tropical storms that feature a low-pressure system, strong winds, and heavy rain. They're usually followed by a storm surge that causes flooding and destruction. While they're all technically classified as "tropical cyclones," storms in the Pacific are called cyclones (or typhoons if they're off the coast of Japan). The same type of storm is called a hurricane when it occurs in the Atlantic or the Gulf of Mexico.

The word *hurricane* comes from the Kalingo people, original inhabitants of the Lesser Antilles, and their god, *Hurican*. They got the word from the Mayan god *Hurakan*, one of their creator gods. The first use of the word *hurricane* occurred in 1667 when a pamphlet entitled "Strange News from Virginia" reported a "dreadful Hurry Cane," during which "the sea swelled 12 foot above the normal height, drowning the whole country before it, with many of the inhabitants, their cattle, and goods."

MEASURING UP

Hurricanes are distinguished by their "eye," which is the calm core of the storm and "eye wall," the winds that rotate out from it. In the 1960s, two meteorologists came up with a rating scale for hurricanes.

The Saffir-Simpson scale rates hurricanes by category:

CATEGORY 1 Winds 74–95 mph

CATEGORY 2 Winds 96–110 mph

CATEGORY 3 Winds 111–130 mph

CATEGORY 4 Winds 131–155 mph

CATEGORY 5 Winds over 155 mph

WHAT'S IN A NAME?

In the past, hurricanes were named either for the name of the saint on whose day the storm hit (for example, the San Felipe Hurricane of 1876 hit Puerto Rico on the feast day of San Felipe) or for the name of the place where the storm made landfall (such as the Galveston storm of 1900).

The naming of cyclones or tropical storms started around 1900 in Australia but lasted only a short time, and then resumed again in the 1940s. The first named hurricane in the U.S. was Hurricane George in 1947. Hurricanes are given names starting with the letter A for the first storm of the year and proceeding through the alphabet. If there are more than twenty-six storms in any given season, they are then named with letters of the Greek alphabet, starting with Alpha.

In 1975, Australia started to alternate the names of storms with male and female names. That practice was adopted worldwide in 1979.

HURRICANE ATLAS

Aptly named "Hurricane Alley," this is the area stretching from North Africa to the Gulf Coast of the U.S., where warm waters provide the right ingredients for monster storms. Florida is the most hurricane-prone place in the U.S. Grand Cayman Island is affected by hurricanes most frequently.

Hurricanes on the Pacific coast of the U.S. and Mexico are rare, but in 1939, a hurricane hit Los Angeles and Santa Barbara. In 1959, a hurricane hit Mexico's Pacific Coast, killing two thousand people.

Not officially hurricanes (because they're not tropical in origin), the "gales" of the North Atlantic can carry winds of up to 120 miles per hour. The gale of 1703 destroyed fifteen warships off the coast of Bristol, England, killing 1,500 seamen. A storm known as "The Big Wind" hit County Galway, Ireland, in 1839, causing hundreds of shipwrecks.

HURRICANE HISTORY

1609: A hurricane that hit a group of colonists leaving England for North America became the basis for Shakespeare's *The Tempest*.

1715: As General Juan Esteban Ubilla left Havana for Spain in 1715, he rounded the coast of Florida with an eleven-ship armada full of gold, silver, porcelain, and silk: cargo worth more than fourteen million pesos. The fleet hit a hurricane, destroying every ship but one. One thousand men were lost, along with the contents of the ships. Four million pesos in gold were recovered by divers six months later.

1780: Estimated wind speeds reached 135 miles per hour during "The Great Storm of 1780," in the Caribbean. It is said to have taken the lives of 22,000 people; 9,000 on the island of Martinique alone. Every home on the island of St. Pierre was destroyed.

1815: "The Great Gale of 1815" is notable in that it struck farther north than most storms. It flooded Providence, Rhode Island, and was the first recorded New England hurricane in 180 years.

1886: The Indianola Hurricane of 1886 destroyed what had been the major port city of Texas.

1900: The Galveston Hurricane of 1900 was chronicled in the book *Isaac's Storm*. It was the deadliest natural disaster in U.S. history, killing more than 6,000 people. The storm pointed out the need for a warning system that would alert residents to oncoming storms so that they could evacuate.

1935: The Labor Day Hurricane of 1935 was the most powerful hurricane to make landfall in the U.S. It whipped through the Florida Keys at speeds of 200 miles per hour. Between 400 and 1,000 people died in the storm.

1938: "The Long Island Express" moved up the East Coast from North Carolina to Long Island in just a few hours, killing 600 people along the way.

1961: Hurricane Hattie devastated Belize to such an extent that the country's capital was moved inland as a result.

1965: Hurricane Betsy, a Category 3 storm, ripped through Florida and southern Louisiana.

1969: With winds of 180 miles per hour and gusts up to 200 miles per hour, Hurricane Camille left 250 people dead from Louisiana to Virginia.

1974: On Christmas Day, Cyclone Tracy hit Darwin, Australia, with winds of 165 miles per hour. Fifty people perished and hundreds more were injured.

1985: The first hurricane to hit Halifax, Nova Scotia, in over 100 years, Hurricane Juan is considered the most damaging storm in Halifax history.

1989: Category 4 Hurricane Hugo destroyed 90 percent of the houses in St. Kitts, Nevis, and St. Croix. In the U.S., it made landfall near Charleston, South Carolina, with winds of 135 mile per hour. It was one of the costliest storms in U.S. history.

1992: Category 5 Hurricane Andrew hit Florida with 165-mile-per-hour winds. Andrew is the second-costliest hurricane in U.S. history.

1999: One of the most devastating hurricanes to hit Central America, Hurricane Mitch hit the coast with winds of 190 miles per hour and triggered mudslides, which left 11,000 people dead.

2005: The costliest natural disaster in U.S. history—and one of the deadliest—Hurricane Katrina took the lives of approximately 1,835 people, and caused over $81 billion in damages.

2011: One of the worst cyclones ever to hit Australia, Cyclone Yasi packed winds stronger than those of Hurricane Katrina.

DECEMBER SURPRISE

The Atlantic hurricane season typically lasts from May through November, peaking in October. In 2005, Hurricane Epsilon arrived in December, becoming only the fifth December hurricane in more than 120 years. Hurricane Lili in 1984 was another, arriving in mid-December.

WASHED ASHORE

Strange things are known to surface and wash ashore during hurricanes.

In 1819 in Mobile, Alabama, a hurricane brought alligators and snapping turtles into the city as the storm surge flooded the streets.

After Hurricane Ike in 2008, a "mystery ship" washed ashore in Alabama. Archeologists determined that it was probably a schooner from the Civil War that had run aground in 1862. Also after Hurricane Ike, a Texas man found an ammunition box washed ashore. The box contained Confederate money, military medals, and diamond earrings.

Six weeks after a cyclone in Pakistan in 2010, a seaside village began to find thousands of smelly mollusks on the beach each night between 6 and 7 p.m.

After Cyclone Yasi hit Queensland, Australia, in 2011, the wreck of the brigantine *Belle* was uncovered. It had sunk in Ramsey Bay in 1880.

In Puerto Rico in 2011, a shark was found in a city street as the result of a storm surge. Tires washed up on a North Carolina beach during the same storm. They had been put into a reef offshore in the 1970s to stave off beach erosion. Meanwhile, a safe washed ashore in Connecticut. The finders called the police and located the safe's owner.

THE UNEXPECTED SIDE EFFECTS OF HURRICANES

In 1988, a huge increase in the number of locusts in Brazil was explained by Hurricane Gilbert, which is said to have blown them across the Atlantic from Africa.

After Hurricane Nora in 1997, sea salt and plankton were found in clouds over the Great Plains of the U.S.

In many hurricanes, sea birds have been blown as many as one hundred miles inland.

After the devastating floods caused by Hurricane Irene in the Mid-Atlantic and the Northeast of the U.S. in 2011, biologists said that the hurricane may have had one unexpected result that will make some people happy. The hurricane created perfect conditions for the cultivation of "magic mushrooms," a hallucinogenic mushroom.

Freakish Phenomena

There may be a perfectly scientific explanation, but that doesn't make it any less strange when it's as dark as midnight at 11 a.m., and you're not in the Land of the Midnight Sun. Through the years, dramatic weather events have caused panic, alarm, and speculation about the end of the world. And then there are those, such as chronicler of all weird weather, Charles Fort, who saw such as evidence of another dimension, which he described as a "super-sargasso" above the clouds from which random things plummet. But sometimes the forces of nature just conspire to throw something really unexpected our way. It keeps things interesting.

Electrifying

At any given moment on our planet, a thousand electrically active thunderstorms are taking place. There are an average of eight million lightning strikes worldwide every day. With all that energy bouncing around, it's no surprise that there are some pretty unusual electrical events that occur. Ordinary lightning is interesting enough in itself, but it's just one end of a very long, electrifying spectrum.

ON THE BALL

Ball lightning is a mischievous (sometimes malicious) glowing ball of light that floats, zigzags, or even bounces off the ground. It's said to be able roll down through chimneys and bounce through a room or even attach to a fingertip. But because it lasts such a brief time, there's not much known about ball lightning, and few have been able to study it.

According to a New Zealand researcher who's an expert in the phenomenon, there have been more than ten thousand reports of ball lightning around the world. While some people decide not to report seeing ball lightning because they think others won't believe them, it has been reported by many famous people including Pliny the Elder, Charlemagne, and Nobel Prize winners in physics Niels Bohr and Pjotr Kapitza.

Ball lightning is typically associated with thunderstorms, but it has also appeared in clear weather, in closed rooms, and even inside airplanes. Some report that the balls seem to move as if directed by an unseen force.

Descriptions of ball lightning vary. In dimension, it can be anywhere from the size of a pea to the size of a house. In color it varies from white to green to orange to violet. In intensity, it typically gives off the same amount of glow as a 100-watt bulb. While ordinary lightning flashes last milliseconds, ball lightning is said to be more slow moving and to hover, drift, and, "skittle" along invisible electrical lines. Some witnesses report that it sometimes makes a hissing sound or explodes before it disappears. Others report that it glides silently, spins, or bounces off objects. Some even say that ball lightning follows them as they move.

Ball lightning is often said to appear close to the ground after a lightning strike. But scientists can't agree on what causes it. Some believe that it's the result of "plasma clouds" charged with particles that glow. Others think that bolts of regular old lightning can form ball lightning when a strike creates vapor that then condenses and mixes with oxygen, releasing chemical energy that forms the ball lightning. Another theory proposes that it's made up of vaporized silicon particles.

What does ball lightning do? Some reports say that it has melted through glass, burned through wood, or set fire to buildings. A lightning researcher was even killed by it in 1753. But usually, it merely surprises those who see it.

BALL LIGHTNING HISTORY

In the 1683 "Great Thunderstorm of Widecombe-in-the-Moor," in England, an eight-foot diameter ball of fire smashed into a church, breaking into two pieces, one of which subsequently disappeared. At the time it was thought to be the work of the Devil, but scientists now believe it was ball lightning.

An 1843 ball-lightning incident was reported by a Parisian tailor. In his story, a ball of fire zoomed down his chimney one night, raced around his living room, and then bounced back into the fireplace and up the chimney, where it exploded. The man described the ball lightning as "the size of a human head." Another person in the room described it as the size of "a cat showing its paws."

In 1907, two men in Burlington, Vermont, saw a "torpedo-shaped" ball of light with "tongue of fire" coming out of it suspended in the air fifty feet above the surrounding buildings. Shortly afterward, it exploded with a tremendous bang.

In 1922, two English women were having tea during a rainstorm when one picked up a knife and saw a miniature ball of light the size of a pea shoot over the tablecloth and then fizzle out.

During a lightning storm in the 1930s, a man reported that he noticed the air above the carpet start to pulsate in the way it does when heat rises from the ground. A ball of light "the size of a pecan" then rose up from the surface over the carpet and attached itself to a woman's fingertip. As lightning flashed outside the room, the ball disappeared.

In 1940, an English man reported seeing a two-foot-wide ball composed of "strings of green light" appear at his feet while gardening. The ball floated up and over the fence, then down the street to a pub where it exploded with a loud bang.

In 1945, an English woman reported a small, bright, bluish-purple ball of fire that rose over her cooking stove and moved past her, emitting a "singed smell" and a rattling sound. When the woman touched it, it burned her hand and her wedding ring felt hot. As it brushed past her, it left a small hole burned in her dress.

In 1966, a man living in Michigan said he witnessed lightning strike a tree, and then watched as three bluish "soccer-ball sized" balls of light came out from behind the tree and started to float around. Several of them came toward the house. One of the balls hit something and exploded, making a sound "like a transformer blowing out." Another just fizzled out.

In 1984, after a plane was hit by lightning, an illuminated ball was said to emerge from the cockpit and into the passenger cabin of the plane before vanishing. This is just one of several incidents of ball lightning in planes through the years.

In 2010, an Australian farmer reported seeing a meteor with a blue tail streaking through the sky. Shortly after, he saw green balls of light bouncing down the slope of a mountain, running along a metal fence, and then exploding.

Before the deadly Joplin tornado of 2011, a woman reported seeing "red fireballs" the size of basketballs bouncing around her yard.

In 2011 in Russia, ball lightning is said to have struck a tourist bus, running from its antennae into its electrical system and exploding. Three of the tires on the bus blew off, but no one on the bus was injured.

WHAT THE SCIENTISTS SAY

In 2010, scientists at the University of Innsbruck in Austria proposed that ball lightning is simply a hallucination that results from an overstimulated brain that's been affected by the electromagnetism of a storm. Through experiments, the scientists found that the magnetic fields focused on the brain can cause subjects to see "luminous discs and lines" that move around. But other scientists say that there's evidence that ball lightning really exists in nature and that the research can't account for the smells and sounds that have been reported with ball lightning, or the varying colors witnesses have seen in the lightning, ranging from blue to green to orange. The consensus seems to be that some, but not all, reports of ball lightning might just be tricks of the brain.

KILLER BOLTS

There's no doubt that lightning can be one of the deadliest forces on the planet. According to the U.S. National Weather Service, an average of seventy-three people die in the U.S. each year as a result of lightning strikes, while many others are injured by it. It is, however, possible to survive single and even multiple lightning strikes. Park Ranger Roy Sullivan of the U.S. National Park Service was struck by lightning seven times between the 1940s and 1970s, sustaining injuries such as singed eyebrows and hair. He later died of completely unrelated causes.

Because they're always outdoors, animals are frequently the victims of lightning strikes. In 1924 in Minnesota, forty-seven heads of cattle were killed by a single strike. In 1939, a single bolt of lightning killed nine hundred sheep in Utah. In Slovakia in 2010, a single bolt killed twenty-seven cows.

The largest single-bolt death recorded for humans happened in Zimbabwe in 1975 when twenty-one people were killed by a bolt that hit a metal shed. Eight were killed in a tobacco barn in North Carolina in 1961.

Florida is the state with the deadliest lightning. Between 1959 and 2003, 425 people in Florida were killed in lightning strikes, while no lightning deaths were recorded in Alaska during the same period.

According to a 1991 study, lightning prefers men to women. Men are more than three times more likely to be killed by a lightning bolt than women are.

CURATIVE POWERS OF LIGHTNING

While reports of lightning injuries are numerous, there are random accounts of lightning strikes *curing* blindness, deafness, and chronic health problems. The claim is that the electrical shock resets the brain or nerves somehow. In 1782, an English clergyman who had been paralyzed was struck by lightning and later recovered from his paralysis. While modern researchers can't prove that occurrence, they have a few modern cases to look to. In 1994, an Oklahoma woman was struck by lightning and later began to feel sensations in her legs, which had long been paralyzed. Days later she was able to walk and eventually fully recovered her mobility. A blind man in Maine was struck by lightning in 1980 and regained his sight. In 2001, an arthritic woman in England was struck by lightning and the next day found that her arthritis was gone.

In a strange case discovered by famed neurological researcher Oliver Sacks, a man was struck by lightning and "became obsessed with music." Even though he had never been musical before, he suddenly heard music in his head, composed it, and taught himself how to play the piano.

LIGHTNING ON THE GOLF COURSE

There's a long-standing theory that most lightning strikes on humans take place on golf courses. In fact, researchers say, most lightning strikes take place on job sites, hitting construction workers on roofs, deck hands on ships, and agricultural workers in fields or nearby sheds.

STRANGE CONDUCTORS

Two women in England were struck and killed by lightning in 1999, and when their bodies were discovered, it appeared that they had been singed around the underwire inside their bras. The story started to circulate that underwire bras can serve as conductors for lightning strikes. Further investigation revealed that the women were hit because they were sitting under a tree. The bras didn't actually attract the lightning, but after they had been hit, the current traveled there.

Can a tongue-piercing act as a conductor? When a Colorado teen was hit by lightning, he felt a burning sensation around his tongue piercing. Experts say that the piercing didn't actually attract the lightning, but rather (in the same manner as the underwire bra) the current ran there after the strike.

NEVER-ENDING LIGHTNING

One of the most bizarre lightning phenomena in the world is the Catatumo Lightning, seen over Lake Maracaibo in Venezuela. Referred to by some as the "never-ending storm," this area sees cloud-to-cloud lightning on 140 to 160 nights of the year, producing flashes up to 280 times per hour, ten hours a night. That adds up to 176,000 electrical discharges per year. These charges are the largest single producer of ozone on the planet. During a severe drought in 2010, the lightning stopped for a four-month period and was the first time in centuries such a thing had happened. The lightning is formed by air from the Andes blowing in and meeting the warm, humid air over the lake and the methane released by decomposition taking place in the marshes. The lightning has been called a "natural lighthouse," which has aided ships through the centuries. It is credited with helping defeat invaders by denying them cover of darkness. The Catatumo Lightning is such a symbol of the area, that it was named a UNESCO National Heritage Site, and locals have also applied to have it made a World Heritage Site.

RED, LIGHT, AND BLUE

From below the clouds, we see flashes of white lightning. But high above the clouds in the upper atmosphere, scientists have discovered incredibly brief flashes of red and blue light in the same storms. They occur over one hundred miles above the Earth's surface, too low for satellites to capture.

Reports of red and blue flashes in the upper atmosphere go back to 1895, but there was no evidence that they really existed. Airline pilots started reporting seeing them in the 1950s and 1960s. The first confirmation of them came in 1989 with research from the U.S. Space Shuttle program. It's only been in very recent years that scientists have gotten pictures of these "luminous phenomena" through cameras mounted on mountaintops. New technology allows the capture of one image per millisecond.

"Sprites, blue jets, and elves" are upper-atmosphere lightning events that seem to happen just milliseconds after a cloud-to-ground lightning strike. Collectively, they've been described as shooting flames, blobs, halos, upward-branching carrots, picket-fence-shaped flashes, and jellyfish shaped. Because they happen so quickly, it's nearly impossible to see them with the naked eye or capture them on film.

Red sprites are described as "tubes" that glow like neon lights. They're often seen with "halos" over them. Blue jets shoot as far as fifty miles up from thunderclouds. Elves are more halo-like in shape and appear around sixty miles up, at the bottom of the ionosphere. Sprites and blue jets develop with a three- to ten-second delay after lightning strikes and move at three times the speed of sound. They can be seen within a sixty-mile radius of the lightning strike.

What causes these fascinating lights? There's no clear answer, but one theory is that vapor forms when lightning strikes, and when it condenses and mixes with oxygen, it releases energy that forms sprites, jets, and elves. Recent research shows that they only occur with very strong lightning strikes, and they're most common with storms that occur over the Great Plains area of the U.S. They're also more common with positive lightning strikes, which occur when positive charges in the clouds result in negatively charged areas on the ground.

COLORED LIGHTNING

Aside from red sprites and blue jets, most lightning is always yellow or white. There have been isolated reports of lightning in other colors, though. In 1976, observers reported seeing lightning in pink, orange, and red with the occasional flash of blue near Northampton, England. A lightning strike in Ontario, Canada, in 1927 is said to have been punctuated by a brilliant blue streak.

Scientists say that these effects are caused by the composition of the atoms in the path of the electric discharge. Blue lightning is said to indicate oncoming hail, while red denotes upcoming rain. Yellow or orange lightning may be caused by dirt particles in the air.

LIGHTNING SHADOWGRAPHS

Can lightning "tattoo" an image onto a surface? There are numerous reports of this phenomenon, said to be "nature's photography." In 1812, a lightning strike, which killed a flock of sheep, is said to have left an imprint of their forms on the grass beneath them. There are several reports of people being struck by lightning under trees that left the imprint of the tree's bark on the bodies.

THUNDERLESS LIGHTNING

Cloud-to-cloud lightning, such as heat lightning, often occurs without thunder, but it's unusual for it to happen with cloud-to-ground lightning. In 1927, a German meteorologist reported a lightning storm that went on for hours in complete silence.

DOWN AND DIRTY

The association between lightning and volcanic eruptions has long puzzled scientists. Going back 150 years, there are reports of lightning bolts appearing over the cones of erupting volcanoes. But the cause of these occurrences has been mysterious. Was the volcano actually producing lightning or just attracting thunderstorms that were naturally occurring? A recent study found that when the ash and rock produced by an eruption collide with each other, they could create a static charge in the same way that ice particles in the clouds produce lightning. Scientists call it a "dirty thunderstorm."

Scientists found that small, quick sparks existed inside the volcano at the time of eruption, but that didn't account for the huge bolts and branch lighting over the erupting cone or explain why some volcanoes make lightning and others don't. One discovery was that the size of the volcano's plume affects the amount of lightning: the higher the plume, the more lightning.

During the 2010 eruption of the Eyjafjallajökull volcano in Iceland and the 2011 eruption of the Chaiten Volcano in Chile, tremendous bolts of lightning were photographed just above the exploding cones. Lightning was also seen in an eruption of Mount Vesuvius in 1944.

EARTHQUAKE LIGHTNING

Just before the Great Kanto Earthquake of 1931, witnesses reported seeing bright, radiating bands of light coming from a single point on the horizon in the manner of a searchlight. Before the 1941 earthquake in Cyprus, witnesses reported seeing a brilliant red globe moving across the island. These phenomena have been called "earthquake lights" or "earthquake lightning." Scientists dismissed these accounts until photographic evidence was recorded during the Matsushiro "earthquake swarm" (a series of earthquakes in a short period of time) in Nagano, Japan, between 1965 and 1967.

Earthquake lights were reported before Peru's Inangahua earthquake in 1968. Before the Sichuan earthquake in China in 2008, people reported seeing a "rainbow cloud" phenomenon in the sky. During the 2011 Christchurch earthquake in New Zealand, earthquake lights were also recorded.

One of the most common theories about earthquake lights is that the friction of rocks striking against each other during the quake causes them. There is speculation that radon is released before earthquakes and that the "glowing clouds" seen at the time of the quakes may just be a reaction of the radon mixing with plasma.

LIGHTNING FLASHES

Central Africa and the Himalayas are considered some of the world's lightning "hot spots," and Darwin, Australia, is called the "storm capital of Australia," with one storm recording more than 1,600 flashes in a single incident. While Florida is the most lightning-prone area of the U.S., researchers estimate that there are twenty-two million lightning strikes across the U.S. each year, the majority of which happen during the summer months. Lightning is most uncommon at the North and South Poles.

There's a popular myth that lightning can't strike the same place twice. This is, in fact, a myth. The Empire State Building in New York is struck on average twenty-five times a year and can even be struck more than once in a single storm.

Especially in high-altitude clouds, rain may form, but evaporate before it hits the ground (this phenomena is called virga, or dry rain). The clouds that form this rain may be thunderclouds and can produce what's called dry lightning. Some statistics say that dry lightning is responsible for 80 percent of the world's forest fires.

There's no rain or thunderstorm, but a sudden flash of blue or purple illuminates the entire sky. It sounds like something from a sci-fi story, but it's been reported at observatories in different parts of the world. One such flash happened in India in 1970 and another in Alaska in 1972. Residents of Southern California reported one in 2010 and in Birmingham, England, in 2011. While such occurrences quickly lead to talk of UFOs, they're more than likely just meteor flashes.

EXTRATERRESTRIAL LIGHTNING

Lightning also exists on other planets. Recent satellite images show lightning storms on Saturn that have lasted for eight months. Lightning on Saturn is one thousand times stronger than it is on Earth. In 2006, NASA scientists were able to photograph a Saturnine lightning storm that was more than two thousand miles wide.

ELECTRIFIED WATER

Can lightning electrify an entire lake? A report from Switzerland in 1672 recounts the story of lightning hitting a lake and instantly killing all the fish, which then floated to the top.

UNDERWATER LIGHTNING

The phenomena of *lapa*, described as "underwater lightning" exists only in the South Pacific. Scientists describe it as a geothermal anomaly that creates flashes and streaks of light far below the surface of the water. It has nothing to do with actual lightning.

ST. ELMO'S FIRE

Yes, it's the name of a 1980s movie, but it's also the name of a luminous phenomenon seen particularly over water during thunderstorms. Green, glowing balls of light seem to spin on the horizon. They can be explained as a corneal discharge, or electrical discharge associated with thunderstorms. St. Elmo's Fire has always been seen as a good omen by sailors. Another mysterious light, the "wil-o'-the-wisp," is often seen bouncing over marshes and boggy areas. It's formed by the emission of methane gas (caused by the decomposition of organic material) as it mixes with oxygen.

Cloudy with a Chance of Weirdness

Look up on any given day and the clouds can tell you whether or not it's going to rain. Look a second time and one of those clouds may look like a submarine, an animal, or even your old Aunt Milly. A third look, however, and a cloud may spit on you, make an unusual sound, shine in the nighttime while releasing insects, or give you the impression Earth is about to be attacked by aliens.

HOLE-PUNCH CLOUDS

On an October day in 2009, reports of a UFO over the city of Moscow started to circulate. An enormous glowing circle, or "halo" as some described it, hovered over the city. With its dark center, it seemed to indicate the presence of a large mothership just on the other side of the clouds. Pictures of the cloud quickly spread around the world, creating much Internet chatter about an alien invasion. But meteorologists were quick to point out an explanation. It was no UFO they said, just a *hole-punch cloud*.

Hole-punch clouds (also called *fallstreak holes*) are a fairly unusual occurrence, but by no means rare. While they can manifest in slightly different ways, they generally look like giant holes punched through the center of an otherwise uniform cloud formation (the Moscow cloud was a little different in that it was a type of inverted hole-punch cloud). They were first reported in the 1940s, not long after the commercial airline industry really took off.

Hole-punch clouds are formed when airplanes fly through sub-freezing, high, thin layers of clouds. When an airplane's wings and propellers move through moisture-filled clouds, the contact can super-cool the water droplets in the cloud. That chilling effect can freeze the water, but since there aren't any particles around to allow ice crystals to form, the frozen water doesn't turn to ice. The water molecules become "sluggish," and then spontaneously produce their own ice crystals without particles. In essence, the airplane creates a "mini-snowstorm" inside a cloud. But instead of falling, the ice stays suspended in the cloud, possibly, researchers say, because the latent heat released in the transformation between water and ice keeps it aloft instead of letting the snow fall.

This snow expands outward uniformly from the place where the plane originally cut through, creating a "hole" in the cloud. The shape that's created in these conditions depends on the position of the plane when it flies through the cloud. If it's level, it may just create a "canal" in the cloud, whereas if the plane was climbing, the hole will look more like a punched hole. The hole can also be "filled in," so there's only a halo-shaped outline in the cloud.

Such was the case in 2010 when more reports of UFOs poured in around Myrtle Beach, South Carolina. There were three nearly identical halo-shaped outlines in the clouds, which hovered over the area for a number of hours. Investigators discovered that planes from a nearby air force base had been flying over the area in formation that day, explaining the uniformity of the hole-punch clouds.

Numerous reports of holes in cloud layers have been reported by ships at sea. In 1963, a boat in the Gulf of Aden reported a huge hole in a thick layer of clouds with what was described as a "tail" of a cloud emerging from the hole. The tail appeared to be rotating in a counterclockwise direction, but wasn't a waterspout or tornado. Scientists believe it was simply an eddy of air.

OTHER "UFO" CLOUDS

Hole-punch clouds aren't the only types of clouds that are frequently mistaken for UFOs. Lenticular or "lens clouds" form when humid air rises and condenses into a cloud. If that air rises through alternate layers of moist and dry air, the clouds will form vertical stacks. There are different types of lenticular clouds, including "cap clouds," which form over mountain peaks and make it appear as though the mountain peak is wearing a beret (or makes it appear as if a UFO is hovering over the mountain).

WHEN CLOUDS ATTACK

Perhaps one of the strangest cloud stories involves a Long Island schoolteacher who, in 1975, reported being "attacked" by a cloud. The man noticed a dark cloud hovering over his house and noticed that it got bigger as he watched it. It enlarged from the size of a basketball into a six-foot oval shape and then into an "abstract, multicurved, vaporous" shape. It moved back and forth over the roof of the house as this happened. The cloud then "seemed to inhale, form pursed lips, and spit water at him."

NIGHT-SHINING CLOUDS

Noctilucent, or "night shining," clouds exist in the upper atmosphere. Scientifically known as *polar mesospheric* clouds, they can only be seen from the ground at latitudes between 70° and 75° north and south, at twilight, during the summer (particularly close to the summer solstice). The shining effect is created by ice crystals in the clouds catching and reflecting the curve of light from the setting sun along the curve of the Earth. Astronauts report being able to see noctilucent clouds glowing blue from outer space.

Noctilucent clouds were first reported in 1885, just after the eruption of the volcano at Krakatoa. While they were previously only seen in higher latitudes, in recent years, noctilucent clouds have been seen as far south as Colorado and Utah in the U.S., a phenomenon scientists attribute to global climate change. While the Earth's surface is getting warmer, the upper atmosphere is getting colder, which is a side effect of carbon dioxide buildup from greenhouse gases.

Nacreous clouds (also called mother-of-pearl clouds) are another type of cloud that is said to "glow." Like noctilucent clouds, they exist in the highest layers of the atmosphere and are seen primarily in winter in northern areas such as Scandinavia, Alaska, and northern Canada. Nacreous clouds can be seen up to two hours after sunset or before dawn. Their colors are described as iridescent.

A 1932 science journal recorded the story of a scientist who saw a "luminous cloud rising majestically in the eastern horizon shining with a uniform, steady, vivid, whitish light." The luminous cloud then released a shower of non-luminous insects.

THE ASIAN BROWN CLOUD

With the advent of satellite photography, meteorologists have been able to observe phenomena from space and get a better sense of global weather trends. That led to the discovery of "the Asian Brown Cloud" a two-mile-thick layer of air pollution that appears over much of Southeast Asia between the months of January and March each year—the months when there are no monsoons to clear the air. While it's not really a cloud (it's more of a layer of the atmosphere), it's full of pollutants from factories, automobile emissions, and particles from wood fires.

Studies estimate that at least half-a-million people a year die as a result of conditions related to the brown cloud. These can include changes in rainfall patterns that cause drought, increased rain, flooding in Australia, retreating glaciers in the Himalayas (which also increases flooding), and decreased crop harvests. The brown cloud is also thought to contribute to the melting of the polar ice caps and rising sea levels. NASA scientists call it a "triple threat" to the environment.

RAINBOW CLOUDS

Rainbows typically span the horizon, but on rare occasions, people report seeing "rainbow clouds," which look like rainbows superimposed on top of wispy clouds. The technical term for this phenomenon is *circumhorizontal arc*. These arcs are caused by light passing through very thin, high-altitude clouds when the sun is high in the horizon and a particular type of ice crystal is present in the cloud. The crystals must be thick and hexagonal, with their faces parallel to the ground. When light hits the vertical surface of these crystals, it will refract in the same way that it would through a prism, thus creating the rainbow effect. Red is the "top" color of rainbow clouds.

Rainbow clouds are typically seen in locations north or south of 55°, but can also be seen at latitudes slightly lower or higher than that around the time of the summer solstice. The possibility of seeing them in Northern Europe is limited. In London, for example, the sun is only high enough in the sky to create them for 140 hours between mid-May and late July. In Los Angeles, that amount increases to 670 hours between late March and late September.

In recent years, urban legends circulated that rainbow clouds signal an impending earthquake. In reality, there's no connection between the two phenomena.

ROLL CLOUDS

A very rare phenomenon, roll clouds look exactly as their name implies. They're horizontal clouds that look like giant rolling pins or tubes. They appear to roll across the horizon along seacoasts, although they have also been seen inland. Roll clouds are formed with downdrafts that spread out before an approaching storm. The drafts sink and hit the surface of the air, causing a wave-like disturbance. Cold flowing air then slides underneath the warm air of the surface and is drawn into an updraft, condensing into a cloud that responds in shape to the wave that was created earlier.

WHAT'S THE STORY, MORNING GLORY?

The "morning glory" is a type of extremely long roll cloud, which is found in coastal areas. While it has occurred in other parts of the world, it is seen most frequently in the eastern sky off the coast of Queensland, Australia, at sunrise, in the months of September, October, and November. Morning glories typically appear in an otherwise cloudless sky and roll toward the shore like a wave at a speed of anywhere between thirty and seventy miles per hour. Morning glories are *solitons*, or solitary waves, meaning that their motion and speed are uniform and do not change. The cloud brings "squall-like" wind or wind shears, but is usually not accompanied by any precipitation. Up to seven morning glories at a time have been reported, arranged in rows. The longest reported morning glory was more than six hundred miles long. Scientists don't have any definitive explanations for this type of cloud.

GRAVITY WAVE CLOUDS

A gravity wave is a vertical air wave caused when air is disrupted and rises up into stable air, and then eventually sinks back down, creating a wave pattern. When this happens, clouds form striking bands or parallel waves.

CLOUD ARCHES

Only a few occurrences have been reported of this strange phenomenon called a *cloud arch*. Each case has been observed by ships fairly far out at sea. Witnesses report that two separate clouds are joined by a narrow arch of clouds between them, often over a long span. There have also been reports of parallel bands of clouds extending between two separate clouds.

LIGHTNING-LESS "NOISY" CLOUDS

While a flash of lightning followed by a clap of thunder is an ordinary occurrence, there have been random reports of noises emanating from clouds completely independent of any storm. In 1931, a Yale professor observed a cloud crossing over Lake Ontario, emitting a continuous "remarkably loud" rumbling noise. There was no lightning to be seen and no rainstorm ever ensued.

In a similarly strange and perhaps less reliable story, in 1814, residents of Agen, France, reported seeing a small white cloud appear that was at first motionless, but then began to spin. Deafening rumbling noises came from the cloud, and a shower of rocks and stones burst from it. The cloud then dissipated.

In 2005, a couple honeymooning in Florida reported watching a single, very low and slow-moving cloud approach. It was the only cloud in the sky that day. As it approached, they heard "a clanging, banging" sound, of the sort you would hear "when a mechanic was working on your car." The cloud "clung to the coastline" as it continued on its course. Hundreds of people along the coast reported hearing the same "noisy cloud" that day, but no explanation for it was ever forthcoming.

HONEYCOMB CLOUDS

A type of cloud that's only been discovered since satellite imaging became available, low-lying "honeycomb clouds" are found over the ocean and distinguished by their interlocking hexagonal cells (hence the name "honeycomb"), which open and close in a synchronized pattern. Scientists say the clouds "communicate" with each other. As each cell rains, a new cell will open up and rain. The cells can even rain in unison. Observers of the clouds' behavior say that the pattern plays out over the course of days and then repeats itself. The clouds are considered to be "self-organizing" and have been compared to fireflies because they can synchronize their rain as fireflies synchronize their flashes.

AGAINST THE WIND

A ship's crew reported an unusual cloud off the coast of West Africa in 1870. The sailors described it as a low-floating, circular cloud that held within it a semicircle divided into four parts. There was a "shaft" dividing the cloud down the center, then "extending out and curving backward." The unique thing about the cloud was that it was travelling against the wind.

Totally Foggy

A whirling mist of clouds seems innocent enough. But as anyone who's ever watched an old movie about Jack the Ripper can tell you, danger can lurk in the fog in the darkness.

Fog is created when warm, humid air is cooled below its dew point (the temperature point water must reach in order to condense). It can happen any time, over any body of water, but it gets particularly thick and persistent in certain places. In the North Atlantic off the Maritime Provinces of Canada, warm air that's following the Gulf Stream meets up with air over the cold Labrador Current, creating a persistent fog, particularly from spring into mid-summer. *Haar* is the name of the fog that lingers over the coast of eastern Scotland in the summer. Parts of coastal Southern California experience what's called the "June Gloom," a foggy period that actually lasts all summer.

LOST AT SEA

While everyone knows about the *Titanic*, the wreck of the *Empress of Ireland* (the worst shipwreck in Canadian history) is an often-forgotten maritime tragedy caused by fog. It was May 1914, just before the outbreak of World War I. The *Empress of Ireland*, a better-than-average ocean liner, was on its way out of Quebec on the St. Lawrence River, headed for England. The *Storstad*, a Norwegian cargo ship, was on its way into Quebec. It was a calm, clear night, when a dense, all-enveloping fog rolled in suddenly, causing near-zero visibility. Being overly cautious (and trying to avoid the shoreline), each ship overcorrected, and the *Storstad* plowed right into the *Empress*. The ship sank within fourteen minutes, taking the lives of 1,012 people.

Modern technology, such as radar, has made sea travel much safer, but seafarers still get lost in the fog, especially on board crafts without navigational equipment. It's not unusual for kayakers to become lost in the fog off Cape Cod, Baja California, or the Pacific Northwest.

IN THE SOUP

One of the most commonly used descriptions of fog is "pea soup," which defines fog that's dense and thick like the soup. Through the centuries, there have been hundreds of literary references to London's pea-soup fog, going back to the thirteenth century. But it was really after the Industrial Revolution that the fog started to take on color as a result of the pollution spewing from the many factories in what was then the heart of the world's economic engine. Starting in the nineteenth century, London fog, in particular, was often described as brown or yellow. The poet T. S. Eliot wrote about "the yellow fog that rubs its back on windowpanes" and "the brown fog of a winter dawn."

Fog was the source of one of the worst environmental disasters in British history: the Great Smog of 1952. Unseasonably cold weather hit southern England that year, and Londoners burnt more coal than usual to heat their homes. With millions of chimneys in the city, that created an unusually high amount of pollution in the air. Combined with the emissions from local factories, the particles from the burning coal left the air full of particulate matter. An especially dense fog rolled in on December 5 of that year just at the same time that the wind died, ensuring that it wouldn't blow away. At times there was nearly complete darkness at midday.

The streets were littered with abandoned cars because their drivers could no longer see to drive. Heathrow Airport shut down. Smog entered buildings through cracks in windows and under doors. People reported that their clothing and even underwear turned black if they stood in the fog too long. The cattle that ingested the fog started to die of black lung.

On the second day of the fog, five hundred people died. The victims were primarily the elderly and people who already suffered from respiratory illnesses such as asthma. Their lungs filled with particulate matter and they suffocated. On the third day, the ambulances had stopped running because they couldn't make it through the fog. That led to more deaths.

When the fog finally lifted on December 9, the death toll had risen to four thousand. One hundred thousand people had been ill. But the after-effects of the fog went on for months. It's estimated now that the fog caused twelve thousand deaths. The shocking death toll led to legislation to control pollution. Parliament passed two Clean Air Acts, one in 1956 and one in 1968.

KILLER FOG

In 1930 in Belgium, a "thermal inversion" fog event killed sixty people in three days. In the Meuse Valley, where several steel and chemical plants were located, heavy pollution always existed, but when it mixed with fog and the wind died down, the toxic brew that usually lifted up and out of the valley had nowhere to go. While many blamed the incident on leaks from the chemical plant, it was actually just chemicals and particulate matter trapped in the fog. The event caused symptoms ranging from burning eyes and throats to nausea and shortness of breath. People even reported skin irritation just from being outdoors. Once again, the elderly and those prone to respiratory illness were the victims. Tests of the air later revealed that thirty chemicals were present in the air at the time.

In 1948, in Donora, Pennsylvania, a thick fog trapped gases from local factories and coal ash from residential furnaces in the air, creating a deadly fog that killed two thousand and caused health problems for six thousand others.

In 1966, extreme smog killed more than 150 people in New York City in a single day. This "killer smog" was never explained, but caused respiratory failure and heart attacks.

In 1973, in Pittsburgh, a four-day-long fog was blamed for fourteen deaths.

In recent years, fog has trapped pollution in some of China's biggest and most populous cities. A particularly thick fog even threatened the cancellation of some of the events in Beijing's 2008 Summer Olympics. The air quality was deemed unsafe for athletes.

RED FOG

"Red fog," or dry fog, has been reported by sailors off the coast of Africa near the Cape Verde Islands for centuries. It's not really a fog at all, but consists of fine dust particles blown from the Sahara. The particles are so fine that it's impossible to sweep them up, and the appearance is that of a dark fog, dense in nature, and capable of causing the same kind of limited visibility created by fog. In 1898, a 1,500-mile-wide patch of red fog was reported in the Cape Verde Islands. The fog was said to have been brought in on a strong breeze, but after it arrived, the air was still. While the fog persisted, the sun and stars were mostly obscured, but some described seeing the sun as either red or "a perfect blue ball." Similar red fogs have been seen around Indonesia, consisting of dust blown in from the Australian deserts.

TULE FOG

Tule fog is a kind of fog that occurs from November through March in the San Joaquin and Central Valley regions of California, near Sacramento. It gets its name from the marsh grass that grows in the area. Tule fog comes in suddenly and is especially dense, reducing visibility to as low as two hundred feet. It's been blamed for numerous traffic accidents along California freeways, causing pileups when one driver is suddenly unable to see and slams on the brakes. In 2007, tule fog was blamed for a one-hundred-car pileup in Fresno that killed two people.

THE "BLASTING" FOG OF CONNECTICUT

Residents of Kensington, Connecticut, reported a strange fog in April of 1758. It was said to appear as thick clouds that then changed to a steam-like appearance when they made contact with buildings. The fog was so hot that people assumed something was on fire and many felt they couldn't breathe. No explanation for the occurrence was ever arrived upon, but people in another town nearby reported having the same experience on that day.

FREEZING FOG

Freezing fog happens when water droplets in the air become super cooled, reaching a temperature of -40°F. While water at that temperature should turn to solid ice, these droplets remain liquid until they find something they can freeze onto. When they do find that object, it will become coated with ice.

Ice fog is a little different. It's composed of tiny ice crystals. Ice fog happens in only the coldest areas of the world because the temperature has to be below 14°F in order for it to form. Ice fog is common in Siberia, especially in the city of Yakutsk (which is also, for the record, the coldest inhabited city in the world).

There's a particular kind of ice fog called a *pogonip*, which only happens in the western U.S. (including Alaska) and Canadian Northwest.

When the air is at nearly 100 percent humidity and the air temperature drops below 32°F, a pogonip can develop. Ice crystals then form in the air. Pogonips are rare in most places, and have been historically seen in Nevada and Alaska in the U.S., and in Greenland and Antarctica. One nineteenth-century pogonip in Nevada lasted for five days.

Nineteenth-century explorers in Greenland reported experiencing pogonips, which caused blisters on their faces and hands. The Native American nations of the West called pogonip the "white death" because they believed the ice crystals could get into their lungs and kill them. In the Jack London story *Smoke Bellew*, a character is actually killed in a pogonip.

"Diamond dust," or ice dust, is described as little "needles of ice" that float in the air creating a fog in areas where the temperature is under -13°F. It's found in Antarctica 316 days out of the year. It has also been seen in northern polar areas and Wisconsin. Astronomers have found that diamond dust is present in Martian ice fog, which is present near the planet's poles.

SEEING THINGS

You've heard of a mirage: a hazy image off in the distance that may look like a city or an object moving on the horizon that only disappears as you get closer to it. A mirage can happen when light travels through air with differing levels of density. The light is refracted (or bent) as it passes through the different levels. When this happens, things can appear well above their actual location, look taller or shorter than they really are, or look compressed, inverted, or multiplied.

Mirages can create a "superior image," which makes the object behind it look higher up than it actually is. Superior images take place when warm air exists above cool air. Sunsets are actually superior mirages because they appear about two minutes after the actual sunset. Superior mirages are caused by cold air beneath warmer air, making them common in polar regions.

A kind of superior mirage, the "looming" mirage commonly happens over cold water or snow when the air above it is warmer. The higher the temperature differential, the higher up the image will seem to loom above the surface of the water or landscape. "Towering" superior mirages appear when the temperature increases more rapidly with height. They're common over large bodies of water in the summer and in polar regions.

Mirages are fairly common on the Great Lakes. The following was reported in 1900: "On Wednesday afternoon people on the lake front at Chicago saw a mirage, in which appeared the sand hills and shrubbery of the Michigan shore. It seemed as if the lake had suddenly grown narrow and that a man could row across in two or three hours. At each end, the shore seemed to run up to the clouds, giving the impression that Michigan was up in the air. The mirage lasted for some time."

In 1977, residents of Grand Haven, Michigan, reported distinctly seeing city lights across Lake Michigan. The closest city is Milwaukee, Wisconsin, which is seventy-five miles away and not normally visible. It turned out they actually did see the lights of Milwaukee, which appeared higher than normal on the horizon because of a superior mirage.

FAIRY LIGHTS

Named for the enchantress in the legends of King Arthur, the Fata Morgana is a superior mirage that appears like a city with towers and turrets hovering on the horizon. The reality behind the Fata Morgana might actually be low hills, beaches, or sea ice. Fata Morgana can also make it seem as though there are multiples of the same landscape feature. Under certain conditions, even flat or smooth surfaces can appear as towers and spires. A Fata Morgana occurs when the temperature increases slowly until it reaches a less-dense air level where it increases more rapidly, and then hits another dense level of air and slows down again. That creates an image that is alternately stretched out and compressed.

A Fata Morgana can be seen on land or at sea, in polar regions, or in deserts. The Fata Morgana was first recorded in the Straits of Messina, between the Italian mainland and Sicily. Through the centuries, sailors in that area have reported seeing "ghost" ships on the horizon or coastlines that aren't really there.

The Fata Morgana is not an unusual site in Fairbanks, Alaska. The towering Alaska Range, visible beyond the city, is distorted, looking taller than it actually is in some places and flatter than it actually is in others. The illusion usually occurs in mid-winter on clear, windless days when the temperature differential between the surrounding valleys and mountains can be as much as thirty degrees. Cool air develops below warm air and when sunlight passes through the air of different temperatures, it creates the distortion.

The Fata Bromosa or "fairy fog" isn't quite as dramatic as the Fata Morgana. It creates more of a fuzzy "wall" or a fog bank hovering over the sea or snow.

The Novaya Zemlya mirage is the name given to a distorted image of the sun above the horizon when it's dark outside. It was reported first near the Russian Arctic island of Novaya Zemlya by an early explorer trying to find the Northeast Passage (around northern Russia to the Pacific), and then again by Antarctic explorer Ernest Shackleton. The Novaya Zemlya mirage is created when the sun's light bends continuously with the curve of the Earth, allowing the light to appear above the horizon.

The *halgerndingar* (Icelandic) or Arctic mirage happens when light passes through uniform temperature variations. This mirage makes it appear as though the entire horizon is elevated. At a certain temperature, the horizon seems to turn upward at the edges, creating a saucer shape.

An inferior mirage is the opposite of a superior mirage. It happens in a circumstance that is the reverse of superior mirages: cool air above a warmer surface. This is the kind of mirage associated with hot pavement and deserts. In an inferior mirage, features on the horizon appear lower than they actually are.

REDWOODS IN THE MIST

While some find the Northern California coastal fog annoying at times, it's a critical resource for the redwoods found in the area. The moisture from the fog keeps them thriving, and they would dehydrate without it. But now the fog is thinning. According to a 2010 report by the National Academy of Sciences, the number of hours of fog in the summer in Northern California has decreased by three hours a day since 1901. That's because the temperature difference between the warm inland air and the cool coastal air has decreased through the years. Scientists say they still don't know if it's because of global climate change.

Dark Days and Light Nights

At around 10 a.m. on a May morning in 1780, the sky in New England turned as black as night—so dark that it was impossible to see. Witnesses in Upstate New York said that the sun was blocked even as it rose that morning. It was the beginning of "New England's Dark Day."

By noon in Hartford, Connecticut, the sky was as dark as midnight. Chickens came in to roost as if it were nighttime. A member of the state legislature, which was meeting at the time, commented, "Either the Day of Judgment is at hand or it is not. If it is not, there is no cause for adjournment. If it is, I wish to be found in the line of my duty. I wish candles to be brought."

This memorable effect was created by a combination of smoke from forest fires in Canada, a thick fog, and ordinary cloud cover. The darkness continued into the night, although light started to return gradually throughout the afternoon. Peak darkness hit Boston at 12:45 p.m. and started to abate at 1:10 p.m. The moon that night was said to glow red. In different places around New England, there were reports of the smell of burning coal in the air. The effect was said to have been experienced all the way into Canada and as far south as New Jersey.

After the Dark Day, the air was notably yellow for a few days, and the sun appeared to be red, a common phenomenon in areas where forest fires have burned.

Similar "dark days" have been reported in the aftermath of volcanic eruptions. When the Indonesian Mount Tambora volcano erupted in 1815, it was said that the sky was pitch dark for three days even three hundred miles away.

In 1857, a darkness descended upon Amsterdam, Holland, for a number of hours and was unrelated to any eclipse. Speculation pointed to air pollution.

A complete darkness descended on Oshkosh, Wisconsin, for eight to ten minutes on March 19, 1886. This, too, was chalked up to air pollution.

THE FALSE DAWN

As spooky as a dark day might be, a "false dawn" is just as unnerving. A British meteorologist reported this rarely seen phenomenon in 1933 when he was driving to London about two hours before dawn. The sky was illuminated with a diffuse light that enabled him to see the details of the surrounding countryside. The light seemed to be coming from the south and east. The phenomenon lasted for twenty minutes, and then it became dark again until the actual dawn several hours later.

This story echoes a similarly strange story told by a Florida man in 1886. He reported waking at 3 a.m. on a moonless night to find the sky lit with an unusually bright greenish light "so bright that you could read by it." The phenomenon lasted for an hour. This same sort of "false dawn," or nighttime luminosity, was reported in 1831 over Europe, except that it was described as a "luminous mist." People said they could read by the light at midnight and it lasted for "a considerable time." In 1931, an English astronomer reported an intense "milky white glow" that covered one-fifth of the sky although the sky was cloudless at the time. Scientists generally attribute these lights to "airglow," a phenomenon associated with solar flares.

A different and very rare phenomenon is the "false sunrise," or dawn sundog, which is an optical illusion caused by light reflecting off ice crystals in cirrus clouds while the sun is low in the horizon and is a type of *parhelion*. ("Mock suns" or sun dogs are not so much a type of weather as they are atmospheric phenomena called *parhelia*. They consist of halos seen around the sun, which sometimes look like additional suns.)

A SEASON SKIPPED

In New England, the year 1816 became known as "the year without a summer" or the "Poverty Year." In June, there were several frosts and snowfalls in Albany, New York, and Maine. Sleet fell in Vermont on June 7–8. There was a freeze in Connecticut in July and ice was still present in rivers as far south as Pennsylvania that month. There was particularly bad weather in western Europe and China as well. The blame for this terrible weather was placed on the abundance of ash in the air from the volcanic eruption of Mount Tambora in Indonesia. As a result of all these weather anomalies, crops in the region yielded only 10 percent of normal averages. After that year, there was a mass migration out of New England.

The weather that year was also said to contribute to a collective malaise, which author Mary Shelley said was part of the reason she wrote *Frankenstein* and *Polidori, the Vampire*.

Not Your Ordinary Weather

Some storms defy categorization.

COBWEB STORMS

In 1832, Charles Darwin was aboard the *HMS Beagle* off the coast of Argentina when he noticed a multitude of gossamer threads stuck to the ship's rigging. On closer observation, he saw that there were thousands of spiders lowering themselves onto the deck. The ship was in the midst of a cobweb storm.

Being the world's premiere naturalist, Darwin was aware of "ballooning spiders," which are spiders that, immediately after birth, propel themselves through the air on gossamer strands. One species is even said to create little "hang gliders" out of their silk. These strands catch the air and can carry the spiderlings to incredible distances. In this instance, Darwin was sixty miles off shore, but airline pilots flying as high as ten thousand feet have reported seeing them as well. In Darwin's case, he noted that that ship had sailed through "clouds" of cobwebs while in the bay of Rio de la Plata.

Wisconsin towns on Lake Michigan reported showers of cobwebs in October 1881. The strands were said to be of great length, some up to sixty feet long, and they fell from great heights. There's an eighteenth-century English account of a cobweb shower that lasted all day long, falling so thick that hunting dogs were covered in it and it had to be scraped from their eyes. A rain of "blue silk" was reported over Naumberg, Germany, in 1665. More believably, cobweb clouds were also reported on the coast of Australia in 1974 and in Santa Clara, California, in 2002.

Spider silk is just part of what scientists refer to as "aerial plankton," or tiny or microscopic organisms that float through the air in the same way that plankton moves through water. For each mile of air, there may be as many as thirty-two million insects and organisms aloft at any time.

While scientific explanations for ballooning spiders and their silk abound, many believe the gossamer threads to be "angel hair," a substance that is said to originate from UFOs.

Spiders in Pakistan created another strange phenomenon after extensive flooding in that country left tens of thousands of people and animals homeless. They covered trees entirely with spider webs, creating "tents" the way caterpillars do, but on a massive scale.

SALT STORMS

Across the world there are numerous areas of salt flats. They can be found in northern Utah near the Great Salt Lake, the Salar de Uyuni region of Bolivia, the Aral Sea in Central Asia, and Lake Orumiyeh in Iran, the largest lake in the Middle East. Salt flats are formed when saltwater lakes dry up, either because of drought or man-made interference such as dam building or diversion of rivers. The strange consequence of this occurrence is a salt storm, where airborne particles of salt form low-lying clouds, which then release the salt in windstorms in nearby areas. The salt particles bring along with them particles of pollutants that had been stored in the lake water. So, the "salt storm" that results carries dangerous toxins both in the air and in the runoff from the salt storms. When salt enters water wells, the water becomes bitter and can no longer be used to feed animals or water crops.

Airborne salt can carry in the wind over three hundred miles. In northern Iraq, environmentalists predict that a salt storm from the dehydration of Lake Orumiyeh could result in the destruction of six to eight cities that are currently home to between four and fourteen million people. The lake, which has shrunk by 60 percent in the past twenty years, was once a popular spot for recreation, and home to flamingos and other migrating birds. Now, it's completely surrounded by a ring of salt. Since the area is already prone to dust storms, the combination of dust and salt means it's only a matter of time before a major salt storm occurs. When that happens, scientists predict nearby cities could be covered in salt.

Salt storms are so toxic that they're linked to throat and lung cancer, infant mortality, and lowered life expectancy. In areas near the Aral Sea in southwest Asia (near Uzbekistan and Tajikistan), residents have reported a large increase in health problems linked to salt storms off the sea, which is now only 10 percent of its original size due to engineering and diversion projects over the past centuries.

Salt storms are a frequent occurrence in Salar de Uyuni, but since the area is sparsely populated, these storms haven't had a major effect on people or towns.

There are records of a salt storm on the Massachusetts coast in 1815, which residents said coated all the buildings within a few miles. Salt was reported to have fallen during a hailstorm in Switzerland in 1870. It was speculated at the time that the salt came from the Mediterranean and had been carried in with the wind.

THUNDERSTONES

Some call them mythical, and the legends of many different cultures around the world mention them. It's been speculated that they're just meteors or parts of meteors. But their make-up is completely earthly. So, are there such things as thunderstones?

An eight-pound stone is said to have fallen in central London during an 1876 storm. There's a report from the English countryside in 1885 describing a stone that fell from the sky in a thunderstorm that killed a sheep. It was said to be the size of a coconut and weighed twelve pounds. It was composed of quartz and had an exterior "shell" that broke off easily. Another English report from the 1880s described a "spherical-shaped" object composed of iron that weighed a little less than a pound. It was found in a hole with char marks around it after a lightning strike. Had it not been for the earthly composition, those who found it would have considered it to be a meteor.

In addition to stones, there are reports of "ax heads" being found embedded in trees after lightning strikes in Italy, the former Prussia, Indonesia, and central Africa. Even stranger are the reports of showers of "flinty stones" reported in Texas, Jamaica, and South Dakota in the nineteenth century. Reports of thunderstones also include details of "fizzing sounds" accompanying them.

In the legends of many Native American nations, lightning is said to create these stones, and possession of one ensures protection from being hit by a bolt. In Norse mythology, thunderstones are said to be thrown to Earth by Thor, the god of thunder.

MISTPOUFFERS

This phenomenon with the funny name has been reported from New York and Scotland to Italy and Australia. Its resonant booming sounds call to mind the boom of cannon fire, distant thunder, or, in more modern times, sonic booms. They occur on clear days in the vicinity of water, sometimes seeming to come from under the water itself. These inexplicable noises go by many names: mistpouffers, lake guns, or fog guns. Near Seneca Lake in Upstate New York, they're called Seneca Guns.

These noises have been known to rattle windows and objects on shelves. The Iroquois explained the noises as the sound of the Great Spirit continuing his work in creating the world. Others speculate that the noises are gases escaping through vents in the Earth's surface or, since they occur near lakes, methane escaping from rotting vegetation. Others think it's the sound of tectonic plates shifting or underwater caves collapsing.

THE AIRPLANE EFFECT

You may remember that airplanes can affect clouds. But can they actually affect weather? Scientists pondering this question studied the weather in areas around airports in order to find out. They found that if the right conditions exist, areas within sixty miles of an airport can receive anywhere from 5 to 15 percent more precipitation.

RAIN-ON-SNOW

A rare phenomenon called rain-on-snow has recently been on the rise in Arctic areas. It happens when a sudden blast of warm weather brings rain or causes snow to melt, and then the water seeps down into the snowpack or forms pools atop frozen soil. When the weather turns cold again, the water freezes. While this all seems harmless enough, rain-on-snow has had a devastating effect on Arctic animals such as reindeer, musk oxen, and caribou. When the water refreezes, it prevents the animals from reaching the tundra on which they graze. Since they can't crack through the ice, they starve. In Canada's Northwest Territories, rain-on-snow has been blamed for the deaths of large herds of musk ox, up to a quarter of the previous population. That in turn has a devastating effect on the Inuit people who depend on the oxen for food and fur for clothing. Rain-on-snow has also been reported in the U.S., Sweden, Russia, and Finland. Another side effect of the phenomenon is that it can cause hazardous flooding when the snow melts.

Weather Wonders

Extreme weather can cause more than inconvenience: it's been known to topple governments, wipe cities from the map, and generally change the course of history. There are places on the globe that are barely habitable due to extreme temperatures, lack of precipitation, and more—where the difference between life and death lays in the next day's forecast.

Superstorms

Sometimes circumstances conspire to create a storm that changes history or rearranges the map. The deadliest storms in history are known by name, and their statistics of carnage are mind-boggling.

THE CHILDREN'S BLIZZARD

Starting in January of 1888, a number of deadly blizzards hit the U.S. One of the most heartbreaking was the Children's Blizzard (or Schoolhouse Blizzard), which pummeled the Midwest, particularly Nebraska and South Dakota. The day started with unusually high temperatures for the time of year, when, in a matter of minutes, the mercury dropped by eighteen degrees and the wind and snow blew in with such force that they created white-out conditions. As the storm came in, teachers in rural areas sent children home from school, not realizing the intensity of the storm. Since it had been so warm in the morning, parents had sent their children to school without heavy coats or wraps, and so the kids headed home in light clothing. With near-zero visibility, many children got lost in the snow. Some parents went out looking for their children and were unable to find them, even though they were just feet away.

During the night, the temperatures fell to double-digits below zero, with a windchill that made it feel like -40°F. The snow had accumulated in four- to five-feet drifts by the time it stopped. In the morning, it was discovered that many children had died of hypothermia in the snow or suffocated under the drifts. On the other hand, some had been able to huddle together or dig themselves into haystacks in order to survive. Some teachers and parents had led students to safety, generally by tying a rope to a shelter and having the children hold on so that they wouldn't get lost.

THE GREAT BLIZZARD OF 1888

Just a few months later, the Great Blizzard (also called the Great White Hurricane) occurred—a storm that usually tops the list of worst winter storms in the U.S. From March 11 to 14, between forty and fifty inches of snow fell from the mid-Atlantic to New England, accumulating in drifts of up to fifty feet. People were trapped in their houses for days. In New York City, fifteen thousand people were stranded on the elevated train. More than four hundred deaths were attributed to the storm. The storm is credited with spurring northeastern cities

to move their elevated train lines underground, leading to the formation of the New York City subway system. Telegraph lines were also moved underground, and better systems for forecasting were developed and put in place.

THE BIG BLOW

Could there be such a thing as a frozen hurricane? The winter storm known as the Big Blow, the Freshwater Fury, the White Hurricane, or the Great Lakes Storm of 1913 was described by those who experienced it as a storm of surprising strength. Nineteen ships and all their sailors were lost on the Great Lakes and most of the wrecks have never been found. Several buildings in Cleveland collapsed under the weight of the snow, which was mixed with sleet and ice. The storm was credited with encouraging cities around the Great Lakes to bury their power lines.

THE IRANIAN BLIZZARD OF 1972

Snow is not uncommon in northwestern Iran (along the border with Turkey), but the blizzard that arrived on February 3, 1972, lasted five days and dumped ten feet of snow in some areas, completely burying several remote villages. Reports from Iranian newspapers at the time said that in two villages, there were no survivors. The death toll from the storm was set at four thousand, making it the deadliest blizzard in history.

Killer Cyclones

Cyclones are the Asian-Pacific equivalent of hurricanes in the Atlantic, Caribbean, and Gulf of Mexico, and they're just as deadly.

THE GREAT BAY OF BENGAL CYCLONES

The Bay of Bengal—which spans the area from Sri Lanka, up the east coast of India, along the coast of Bangladesh to the west coast of Myanmar (Burma)—is the most cyclone-prone area in the world. Dating back to the 1500s, there are impossible-to-confirm reports of cyclones that are said to have killed hundreds of thousands of people.

In the twentieth century, seven of the nine deadliest weather events in the world happened in Bangladesh.

A cyclone that hit the Indian British colonial city of Calcutta (now Kolkata) in 1737 was said to have taken the lives of three hundred thousand people, but that figure is doubtful, as the population of the city was less than twenty thousand at the time. Modern researchers attribute this mistake to a typo made in an English magazine article published in 1738 since site reports from British officials who were present calculated the figure as three thousand dead. An article in the *Gentleman's Magazine*, in June 1738, described the storm this way: "On September 30, last happened a furious Hurricane in the Bay of Bengal, attended with a very heavy Rain which raised 15 Inches of Water in six Hours, and a violent Earthquake, which threw down [an] abundance of Houses; and as the Storm reached 60 Leagues [180 miles] up the River Ganges, it is computed that 20,000 Ships, Barks, Sloops, Boats, and Canoes have been cast away. A prodigious Quantity of Cattle of all Sorts, a great many Tygers, and several Rhinoceroses were drowned; even a great many Caymans [crocodiles] were stifled by the furious Agitation of the waters, and an innumerable Quantity of Birds was beat down into the River by the Storm. Two English ships of 500 Tons were thrown into a Village above 200 Fathom [more than 1,000 feet] from the bed of the River Ganges, broke to Pieces, and all the People drowned pellmell among the Inhabitants and Cattle." Modern researchers see no evidence that an earthquake occurred at the time and believe that observers only thought there had been an earthquake because of the destructive power of the wind and water.

In 1864, the city of Calcutta, which had by then grown to a population of over forty-five thousand, was once again hit with a cyclone. While the destruction of the storm surge was not as strong as in the 1737 cyclone, the death toll eventually reached sixty thousand. With unburied bodies lying in the streets and contaminated food and water, cholera ripped through the area wiping out the populations of entire villages in the region in the following months.

Just twelve years later in 1876, the Bengal Cyclone, which hit an area east of Calcutta, is thought to have taken the lives of two hundred thousand when flooding from the Meghna River overtook settlements along its banks.

Considered one of worst natural disasters in modern history, the Great Bhola Cyclone pushed across the Bay of Bengal on November 11, 1970, as a Category 3 storm with winds of approximately ninety miles per hour. It made landfall near Chittagong in what was then East Pakistan (now Bangladesh). The wind demolished everything in its path on thirteen islands in the bay, while the storm surge swept over the islands, drowning all the inhabitants and their livestock. Forty-five thousand fishermen drowned. In the end more than three hundred thousand people lost their lives, making this the deadliest cyclone in recorded history. The government's failure to provide sufficient warning about the storm, and its slow response after it took place, caused great civil unrest in the country and is considered to be the biggest contributing factor in east Pakistan's eventual split from west Pakistan, which formed the nation of Bangladesh.

TERRIBLE TYPHOONS

Like hurricanes, typhoons are tropical cyclones consisting of a low-pressure center, high humidity, warm surface temperatures over water, vertical wind shear, and atmospheric instability. Typhoons are found in the northwest Pacific Ocean, typically affecting eighteen countries, including China, Japan, the Philippines, and Taiwan.

The worst typhoon in modern history was Super Typhoon Nina which hit the Philippine Sea in 1975. Winds reached 115 miles per hour over Taiwan. But the major damage of the storm took place on mainland China where the intense rainfall caused the failure of the Banqiao Dam. More than 170,000 drowned in the resulting flooding.

Storms that Changed History

Who knows what the world would look like today without these storms!

THE KAMIKAZE

While typhoons cause destruction and tragedy, they also turned out to bring salvation on more than one occasion in thirteenth-century Japan.

In 1274, a Mongol war fleet dispatched by Kublai Khan closed in on Japan. Records indicate that there were between fifteen thousand soldiers and five hundred to nine hundred ships in the fleet at anchor in Hakata Bay near Kyushu Province. The battle on land had been going well for the invaders when a storm approached. The commander ordered troops back to the ships so that they wouldn't be stranded on Japanese soil. Overnight, a typhoon developed and more than two hundred ships were lost. The Japanese, in their nimble, small boats, were then able to outmaneuver the Mongols and drive them from their shores. The typhoon was hailed as the *kamikaze*, or divine wind, that drove out the invaders.

Unbelievably, in 1281, the Mongol forces tried again. This time, the Mongols are said to have brought a fleet of more than four thousand ships manned by 140,000 warriors (modern historians compare the scale of this force to the D-Day invasion of Normandy by the Allied troops). The fleet was unable to find a place to make landfall so they stayed on their boats, depleting their supplies until a typhoon hit and destroyed the fleet. That was the last attempted naval assault on Japan by the Mongols.

At the time, the Japanese believed that the god Raijin had protected them with the wind and driven the invaders away. During World War II, the Japanese army adopted the term *kamikaze* to describe their suicide pilots, whom they also hoped would thwart their enemies.

WARRING WINDS

The kamikaze isn't the only wind that is said to have kept invaders at bay. During the Spanish Armada's attack on England during the reign of Elizabeth I, the storm that destroyed the invading fleet became known as the "Protestant Wind." This weather event ensured that England stayed Protestant rather than converting to Catholicism, as it would have done under the Spanish. The term "Protestant Wind" was also used to describe the favorable wind conditions that helped the Dutch forces of William of Orange reach England and depose England's last Catholic monarch, James II.

THE SAN MATEO STORM

The French were also thwarted from the their imperial aspirations in Florida by a major weather event. It was 1565, and the Spanish had just established a fort at St. Augustine. The French had established a fort farther up the coast at Fort Carolina (near what is now Jacksonville). The commander of the French unit decided to move upon St. Augustine by sea, despite the warnings of his top advisor, who predicted a storm. A hurricane struck, and most of the French fleet was destroyed, while those men and ships that did survive were grounded just south of St. Augustine. At the same time, Spanish troops were moving upon the French fort by land. They found it virtually unmanned, and were able to easily destroy it. They then came upon the survivors of the storm and handily defeated them, thus ending any French claim on Florida. The Spanish referred to the hurricane as the San Mateo Storm because it fell just after the feast or saint's day of San Mateo.

THE GOOD NEWS ON REALLY BAD STORMS

While the death tolls in these historic storms are staggering, researchers have recently uncovered some good news. While extreme weather events are becoming more frequent, the death tolls from them have dropped 98 percent since the 1920s.

The study found that deaths due to weather events accounted for less than one-tenth of the deaths per decade, following other causes such as car accidents and wars. While the study didn't point directly to the reason for this precipitous drop, other studies have pointed to better forecasting, due in part to lessons learned from deadly storms. The Galveston Storm of 1900, in particular, is credited with improving the U.S. weather alert system.

Weather Extremes

The hottest hots, the coldest colds, the wettest and the driest. These weather events are anything but average.

HAVING A HEAT WAVE

Few people know that heat waves are the deadliest weather phenomena. Excessive heat can cause heatstroke (also called *hyperthermia*), which is when an individual's body is unable to dissipate heat.

The worst heat wave in recent history hit Europe in the summer of 2003, with record temperatures for the northern hemisphere occurring during August of that year. Much higher than average temperatures were experienced in France, Germany, Italy, and the UK, which experienced its first ever 100°F temperature. More than fourteen thousand people died from the heat in France alone, mostly elderly people who suffered from heatstroke. One of the major problems was that the temperature didn't drop at night. In Paris, five thousand deaths were attributed to high nighttime temperatures. The cumulative death toll for the heat wave was thirty-five thousand across Europe.

In 2010, a heat wave in Russia produced the highest temperatures in the area since 1870 and claimed the lives of fifty-five thousand people. The temperature in Moscow reached 101°F during the day and 77°F at night.

According to the U.S. National Oceanic and Atmospheric Administration, 2010 tied with 2005 as the warmest year since temperature records started in 1880.

SOME LIKE IT HOT

Dallol, Ethiopia, is the hottest inhabited place on Earth with an average temperature of 93.9°F over thirty years of record keeping. Ahwaz, Iran, hits the highest daily maximums, averaging 115°F.

The hottest temperature ever reported was 136°F in El Azizia, Libya, in September of 1922, but the record hasn't been verified. The hottest U.S. temperature, 134°F, was recorded in Death Valley, California, in July of 1913.

Marble Bar, Australia, holds the record for the most days over 100 degrees. For 160 consecutive days in 1924, the town endured 100-plus degree temps. The town's average high is usually over 100 degrees in the months of January through March and November and December.

Temperatures typically drop at night, but in 1995, on the hottest July day on record in Phoenix, Arizona, the daytime temperature of 121°F dipped to 105°F at 9 p.m., and then spiked back up to 114°F at 11 p.m.

The fastest temperature rise ever was recorded in Spearfish, South Dakota, where the temperature rose 49 degrees within two minutes in January of 1943. The change in the weather was attributed to a gust of katabatic wind.

In July 2011, 3,709 high-temperature records were broken in the U.S.

In June 2011, Oman recorded the world's highest minimum temperature ever at 107.1°F.

Earth has warmed an average of 1.4 degrees since the Industrial Revolution.

COLD AS ICE

The coldest place on earth is Vostok, Antarctica, which recorded a temperature of -128.56°F on July 21, 1983. Vostok is a Russian research station, which is occupied by international scientists for a few months each year.

The coldest inhabited city in the world is Yakutsk in eastern Siberia (population 270,000). January is the coldest month, with average highs of nearly -40°F. Yakutsk is also the largest city in the world built on permafrost, a soil that is a compact combination of sand and ice.

Oymyakon, Siberia, not far from Yakutsk, is the location of the lowest temperature ever recorded: –128.6 F.

International Falls, Minnesota, which holds the coveted title "Icebox of the Nation," claims to be the coldest spot in the U.S., but Fraser, Colorado, disputes the claim. Records indicate they're both wrong: the title belongs to Stanley, Idaho, which has a record low temperature for the lower forty-eight states of -52.6°F, plus the highest number of cold days during a twenty-year period. The lowest temperature recorded in the U.S., including Alaska, was set, not surprisingly, in Alaska. The mercury at Prospect Creek, Alaska, hit -80°F in 1971.

The coldest temperature ever recorded in North America is -83.02°F at Snag, Yukon Territory, in 1947.

The fastest temperature drop was within one hundred miles of the fastest temperature rise (page 146) on the same day in January of 1943. The temperature dropped 45 degrees in fifteen minutes, from 60°F to 13°F, due to katabatic winds.

The most enduring period of cold was the "Little Ice Age" in Europe, which extended from 1150 to 1460, with recurrences in the sixteenth and seventeenth centuries. During those periods, warmth-loving crops died, causing widespread famine and starvation. This extended cold wave is believed to be one of the key factors in the decline of the Viking civilization and for civil unrest in England and France.

MEGADROUGHTS

The word *megadrought* describes droughts that last more than two decades. The longest drought in modern history has persisted from 1965 to the present in the Sahel region of central Africa. The cause of the drought is disputed. Some say it's due to land overuse while others point to global climate change. The drought is responsible for the death and displacement of millions of people, internecine warfare, lakes drying up, and species becoming extinct.

A huge drought rewrote Native American history in the thirteenth century when the longstanding Puebloan culture of the Southwest came to an end. Thousands of people are assumed to have died, been displaced, or absorbed into other groups farther south.

The worst drought in history took place in China between 1876 and 1879, when nine provinces received almost no rain, and crops failed over an area of six hundred thousand miles. Nine million people are said to have died.

Between 1932 and 1937, in the Great Plains of the United States, the "Dust Bowl" drought was caused by overuse of the land in an area that had once been grassland. Soil dried up and was carried in dust storms and "black blizzards" (see page 65).

In the Australian grain belt, droughts and wildfires in 2010 caused a strange phenomenon: the spontaneous combustion of haystacks. More than four hundred spontaneous hayshed fires were reported in New South Wales. Scientists say that because the haystacks, usually consisting of wheat and barley, get wet during summer rain, they become hosts to microbial growth (these types of grains are also relatively high in natural sugar content). During the droughts, they dry out, and the combination of dry material and bacteria generates heat and spontaneously combusts.

WORLD'S WETTEST PLACES

What's the world's wettest place? Mount Waialeale in Kauai, Hawaii, often claims the title. It rains between 335 and 360 days a year, averaging out to approximately 460 inches a year based on thirty years of records. But Lloro, Colombia—which receives its rain primarily during a two-month rainy season—gets more in fewer days. Over thirty years of records, the town has averaged 523 inches a year. Also in the competition is Mawsynram, India, which has averaged 467 inches of rain a year in thirty-eight years of records.

Australia's rainiest spot is Mount Bellenden Ker, which averages 310 inches a year, while the rainiest spot in Europe is Crkvica, Bosnia-Herzegovina, which receives an average of 183 inches a year. That's nothing compared to North America's rainiest spot, Henderson Lake, British Columbia, which gets about 256 inches a year, or Africa's rainiest spot, Debundscha, Cameroon, which gets about 405 inches a year.

Cherrapunji, India, holds the record for the most rain in one year and in one calendar month. In 1860 to 1861, the area received 1,042 inches of rain. Over 30 feet of that total occurred in July 1861, making it the rainiest month in history.

Hilo, Hawaii, is the wettest city in the U.S. and Hawaii is the wettest state.

The most rain to fall during a twenty-four-hour period fell in 1966 in Foc Foc, Reunion Islands, when Tropical Cyclone Denise dumped 71.9 inches in one day.

The most rain to fall in a single minute fell in Unionville, Maryland, in 1956 (1.56 inches). Twelve inches fell in a single hour in Holt, Missouri, in 1947.

Freaky Weather Facts

The weather can be strange, but even stranger is the effect it can often have on humans.

CLOUDY WITH A CHANCE OF REDHEADS

The cold, cloudy skies of northern Europe are responsible for . . . red hair? Researchers say that residents of Scotland and Ireland are more likely to have red hair and fair skin because of the climate they live

in. As humans migrated northward, they adapted by developing genetic traits suitable to their new environment, including hair and skin that would have been too fair for sunny climates.

A 2011 report by British scientists also revealed that people who live in northern climates have bigger brains than those who live farther south. The size of the brain, however, doesn't appear to translate to greater intelligence. The study proposed that those who live farther north need bigger brains to process the additional visual information required to live in places with dimmer light.

METEOROLOGICAL MIGRAINES, MEMORIES, AND MOODS

According to a 2011 study at Harvard University, changes in weather can actually cause migraines for those who are already prone to having them. With every nine-degree rise in temperature, migraine sufferers are more than 7 percent more likely to get headaches. They're less likely to get headaches with drops in barometric pressure.

In 2009, Australian researchers discovered that gloomy weather actually improves the memory function of the brain. In the study, subjects were three times as likely to remember the arrangement of items on a counter when the weather was bad than when it was sunny. In sunny weather, the researchers found, people were overconfident about their memories.

A 2010 German study revealed that bad weather doesn't necessarily produce bad moods. Researchers found that all of their subjects fell into one of four categories: people who are unaffected by the weather, people who love summer, people who hate summer, and people who love rain.

There are some who believe that weather on the sun can affect humans. A study by the *New Scientist* in 2008 made a link between the pineal gland of the brain, which is sensitive to magnetism, and solar flares on the sun, which produce larger-than-normal amounts of magnetized particles. The study found that during periods of increased geomagnetic activity, depression and suicide rose. The authors of the study speculate that solar flares can "desynchronize circadian rhythms and melatonin production" in the brain.

ANOTHER REASON NOT TO LIKE SNOW

A 2010 report confirmed that shoveling snow can lead to heart attacks. The report found a correlation between heavy snowfalls and a 22 percent increase in heart attacks in the northeastern U.S.

Ponderous Predictions

Perhaps as strange as the weather itself are the indicators that farmers, sailors, and others have long used to predict it. We've all heard the old adage, "Red sky at night, sailors delight; red sky at morning, sailors take warning." But did you know that you can gauge the temperature by counting cricket chirps? Drawn from farmer's almanacs and conventional wisdom, the following advice outlines weird ways to predict the weather.

THE GROUNDHOG

The most famous of the weather-predicting animals is the groundhog. The current-day superstition is this: Upon exiting his hole on February 2, the groundhog will either see his shadow or not. If he does, winter will persist for another six weeks. If not, winter will soon end.

This tradition came to North American with German immigrants. They adapted it from a European folk belief dating back to pagan times when February 2 was a fertility celebration and thus associated with the coming of spring.

CRICKETS

Count the number of chirps within a fifteen-second interval and then add thirty-seven to that number to get the current temperature in Fahrenheit. To find out what the temperature is in Celsius, count for twenty-five seconds, total your count, divide it by three, and then add four.

WOOLY WORMS

Wooly worms, also called wooly bears, are the caterpillar stage of the tiger moth. Their surface is covered with wooly stripes in alternating bands of black and brown. For hundreds of years, farmers have checked the wooly worm population in the fall to determine if a winter will be harsh or mild. If the brown bands on the wooly worms are wide, the saying goes, winter will be mild. If they're narrow, it's going to be a rough winter. Solid brown wooly worms mean a mild winter is coming; solid black means it will be very bad. According to a study by the American Museum of Natural History in the 1950s, the wooly worm method of predicting winter is 80 percent accurate.

NUTS AND CORN

The thickness of acorns or nutshells and of cornhusks is also said to determine the severity of winter. If they're thin, winter will be mild and vice versa.

PERSIMMON SEEDS

One of the stranger methods of weather prediction has to do with the seeds of a persimmon. The inside of a persimmon seed contains a filament that, depending on its shape and size, can look like a spoon or a shovel. If the majority of the filaments are spoon-shaped, it's said the winter will be mild. If more shovel-shapes are found, expect a lot of snow. There's been absolutely no scientific proof to back up this method of prediction.

BIRDS

Birds are thought to be keenly attuned to the weather. In advance of a hurricane or severe storm, they disappear. Scientists believe that this is due to their sensitivity to barometric pressure.

COWS

If cows are lying down in a field, farmers say rain is on its way.

LEEKS

In England, leeks were long thought to ward off lightning. When thatched roofs were the norm, people grew leeks on the roofs to protect the home from lightning strikes.

UNMARRIED WOMEN

During droughts in Bihar Province of India, farmers send their daughters into the fields to plow. This is thought to shame the weather gods into sending rain.

ST. SWITHIN'S DAY

In England, there's long been a folk belief associated with July 15, St. Swithin's Day: "St. Swithin's Day, if it does rain, full forty days it will remain. St. Swithin's Day, if it be fair, for forty days, t'will rain no more." Studies attempting to prove the legend have fallen flat. There doesn't seem to be any consistent weather pattern for the forty days following St. Swithin's Day.

WEATHER PROVERBS

When stars shine clear and bright, we will have a very cold night.

The sharper the blast, the sooner 'tis past.

If bees stay at home, rain will soon come. If they fly away, fine will be the day.

When the swallow's nest is high, summer is dry. When the swallow's nest is low, you can safely reap and sow.

A rainbow afternoon, good weather coming soon. A rainbow in the morning, is the shepherd's warning. A rainbow at night is the shepherd's delight.

When the chairs squeak, it's of rain they speak.

Index